T0140016

Studies in Big Data

Volume 123

The series "Studies in Big Data" (SBD) publishes new developments and advances in the various areas of Big Data- quickly and with a high quality. The intent is to cover the theory, research, development, and applications of Big Data, as embedded in the fields of engineering, computer science, physics, economics and life sciences. The books of the series refer to the analysis and understanding of large, complex, and/or distributed data sets generated from recent digital sources coming from sensors or other physical instruments as well as simulations, crowd sourcing, social networks or other internet transactions, such as emails or video click streams and other. The series contains monographs, lecture notes and edited volumes in Big Data spanning the areas of computational intelligence including neural networks, evolutionary computation, soft computing, fuzzy systems, as well as artificial intelligence, data mining, modern statistics and Operations research, as well as self-organizing systems. Of particular value to both the contributors and the readership are the short publication timeframe and the world-wide distribution, which enable both wide and rapid dissemination of research output.

The books of this series are reviewed in a single blind peer review process.

Indexed by SCOPUS, EI Compendex, SCIMAGO and zbMATH.

All books published in the series are submitted for consideration in Web of Science.

Aboul Ella Hassanien · Ashraf Darwish ·
Mohamed Torky
Editors

The Future of Metaverse in the Virtual Era and Physical World

Editors
Aboul Ella Hassanien
Faculty of Computers and Artificial Intelligence
Cairo University
Cairo, Egypt

Mohamed Torky
Culture and Science City
Higher Institute of Computer Science and Information Systems
Cairo, Egypt

Faculty of Artificial Intelligence
Egyptian Russian University (ERU)
Cairo, Egypt

Ashraf Darwish
Faculty of Science
Helwan University
Cairo, Egypt

Egyptian Chinese University (ECU)
Cairo, Egypt

ISSN 2197-6503 ISSN 2197-6511 (electronic)
Studies in Big Data
ISBN 978-3-031-29134-0 ISBN 978-3-031-29132-6 (eBook)
https://doi.org/10.1007/978-3-031-29132-6

This Springer imprint is published by the registered company Springer Nature Switzerland AG
The registered company address is: Gewerbestrasse 11, 6330 Cham, Switzerland

Preface

The notion of the Metaverse is profound. Originally used to describe Stephenson's vision of a fully immersive 3D virtual world in 1992, the phrase has since expanded to describe the real-world objects, characters, interfaces, and networks that create and interact with such worlds. The term Metaverse, a combination of the words meta and universe, describes a virtual universe in which avatars participate in political, economic, social, and cultural activities.

There is a history behind the rise of the Metaverse, and it involves several patterns in the exponential increase of technology's capacity and performance. As a robust Metaverse takes shape, it will influence the evolution of other fields of technology that might at first glance seem unrelated to the Internet. It's crucial to bridge the gap between the current metaverse's features, which are mostly Web 2.0 traits, and the forthcoming Web 3.0 traits.

There are key ways in which the Metaverse is apart from Augmented Reality and Virtual Reality. Metaverse stands out as a service with more long-term, meaningful content than Virtual Reality, which is why it's getting so much attention in the research world. In order to strengthen social meaning, it is crucial that the Metaverse provides a scalable environment that can support a large number of users.

The potential for the metaverse to greatly increase consumers' access to the market from emerging and frontier economies is quite exciting. Several technologies, such as always-on mobile connectivity, Artificial Intelligence, the Internet of Things, and blockchain technology with the use of virtual currency to bridge the gap between the virtual and the real worlds, are strengthening the metaverse.

In light of the currently available studies focusing on the metaverse, this field of study is still in its infancy. Consequently, more studies should be searched in the metaverse and its applications. Metaverse growth is still mostly uncharted by academics. As technology advances, it's possible that more of the Metaverse's capabilities will become available, and the virtual worlds will continue to get closer and closer to being a true universe. This book is an attempt to build a bridge between many cutting-edge fields that may help the metaverse advance. This book's secondary objective is to examine the most recent developments in the fields of the user interface, implementation, and application inside the Metaverse.

The contents of this book are divided into three parts. Part I is devoted to the Metaverse concept, history, origin, and applications. Part II is concerned with the role of emerging technologies such as Artificial Intelligence, the Internet of Things, and digital twins. Part III presents the role and use of cybersecurity such as blockchain, privacy, and security technologies.

The authors of this book would like to thank all authors of this book for their contributions and research studies.

Cairo, Egypt

Prof. Aboul Ella Hassanien
Prof. Ashraf Darwish
Assist. Prof. Mohamed Torky

Contents

Emerging Technologies in Metaverse World

Cybersecurity in Metaverse

Metaverse and its Applications

Development of Cognitive Intelligent Mechanism for Sustainability of Bigdata: A Future Shape of Metaverse

Debabrata Datta and Madhubrata Bhattacharya

Abstract Metaverse is proposed as a mapping of virtual world to the physical real world using cognitive intelligence hybridized with artificial intelligence. Cognitive intelligence will be used to develop a meta brain of virtual world through mapping of functional state spaces and artificial intelligence will be shaping the intelligence of various functions due to which it will be possible to have transition from one state to other. Conceptual space of metaverse like other conventional spaces such as Hilbert space, Sobolov space and Euclidian space will be handling big data with full proof security like Blockchain. Cognitive intelligence and Artificial intelligence both when combined together will develop a general intelligence known as Artificial General Intelligence. Sustainability of big data may be an issue for future shape of metaverse but that will be achieved using Artificial General Intelligence. Human centric functional modeling using cognitive intelligence is described in detail in the chapter. Various devices that can be used for future shape of metaverse are demonstrated. Virtual reality, augmented reality, extended reality and mixed reality are backbones of future shape of metaverse and all these concepts are presented detail in the chapter.

Keywords Cognitive intelligence · Artificial intelligence · Metaverse · Big data · Virtual reality

1 Introduction

Industry of various sectors like healthcare, medical, education, agriculture, electronics and other engineering and scientific fields needs an innovative automation for handling big data with its sustainability property. The fulfilment of this objective is

D. Datta (✉)
Department of Information Technology, Heritage Institute of Technology, Kolkata, India
e-mail: debabrata.datta@heritageit.edu

M. Bhattacharya
Department of Physics, The Heritage College, Chowbaga Road, AnandaPur, Kolkata 700107, India
e-mail: madhubrata.bhattacharya@thc.edu.in

A. E. Hassanien et al. (eds.), *The Future of Metaverse in the Virtual Era and Physical World*, Studies in Big Data 123, https://doi.org/10.1007/978-3-031-29132-6_1

only possible when one understands the fundamentals of metaverse, a combination of meta and universe. Conceptually, the metaverse is a future generation technology that goes beyond our current universe. In 1992, Neal Stephenson used the word 'Metaverse' in Snow Crash [1], his science fiction novel, where humans in the physical world live in a parallel virtual world known as metaversethrough digital representation of users. But the technology was not fruitful enough to reinforce his thought and that is why it did not take any shape. In 2003, the virtual game 'Second Life' was developed, where a virtual world was introduced with all its social elements. Artificial intelligence (AI), Blockchain (BC), Extended Reality (XR), Virtual Reality (VR), Cognitive Intelligence (CI), Brain Informatics (BI) and Data Science (DS) have developed a new generation of technology. Among these methods AI, BI, CI and DS manifested great significance of big data processing to increase the immersive environment and empower the avatars to possess human like intelligence (e.g., Humanoid). Many researchers have done lots of studies on AI which are mainly focused on physical world. But for metaversethe AI should make people enable to navigatevirtual environment and real world with augmented reality.

Looking at the limitations of offline social activity during period of Covid-19 (2019–2020), the year 2021 is reported as the first year of the metaverse. In October 2021, Mark Zuckerberg rebranded Facebook as Meta because this social media generates a substantial amount of new data. Metadata is defined as a set of data that generates another set of new data. Mathematically, metadata if belongs to a universe, then it will create metauniverse or metaverse and following this concept of metaverse our life will change drastically. Big companies have started investing to create the metaverse technology. For an example "Activision Blizzard", a video game company is bought by Microsoft for $68.7 [2] billion for gaming expansion in the metaverse. In recent years an outburst of metaverse happened in 3D gaming. Virtual world has been built more creatively using improved big data storage, graphic processing unit-GPU and wireless network. AI has been deployed in different domain like neural interface, natural language processing etc. Layered architecture of metaverse can be organized into following seven layers [3] (see Fig. 1).

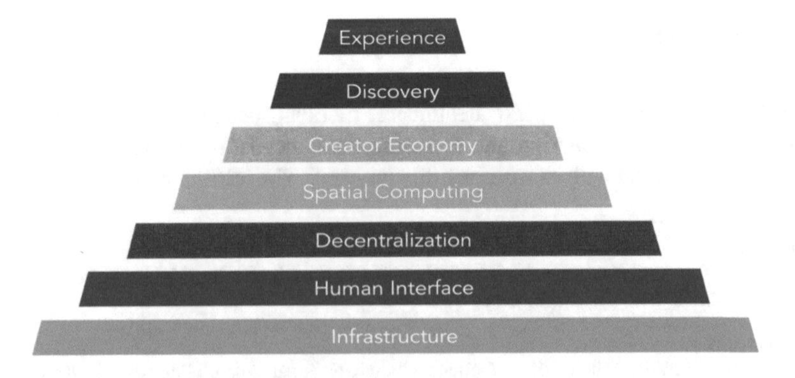

Fig. 1 Seven layers of metaverse platform

(a) Infrastructure: WiFi, data center, clould, CPUs, GPUs, 5G, 6G.
(b) Human Interface: smart watch, smart phone, smart glasses, display device (mounted on head), acoustic device.
(c) Decentralization: AI agents, block chain, edge computing.
(d) Spatial Computing: Virtual Reality (VR), Augmented Reality (AR), Extended Reality (XR), multitasking.
(e) Creator economy: E commerce, design tools, asset market.
(f) Discovery: avatar, virtual stores, advertising networks.
(g) Experience: learning, gaming, shopping, working.

Technically metaverse can be illustrated by a combination of information technology and big data. All the information and activities of the user in the metaverse are taken down as data. Number of users if increasesthen huge amount of data generation takes place which further establishes a big data network. With an appropriate route of security such as blockchain one can achieve the sustainability of big data and subsequently we can create a metaverse that can shape the future world. However, sustainability of big data is an issue for creation of metaverse. Conventional method of handling big data is not sufficient for its sustainability and for that reason an innovative intelligent technique that is combination of AI, BI and CI is proposing for creation of future metaverse. The data grows continuously and are processed by data processing technology with appropriate security using blockchain technology [4, 5]. Big data is defined as the set of five V's explaining each V as (a) volume of data, (b) velocity (rate at which data generates), (c) variety (text, image, numerical), (d) veracity (uncertainty of data) and (e) value (worth to analyze). Therefore, big data cannot be analyzed by conventional data processing tools due to its complexity, massive structureand diversity [6]. Analysis of big data generally needs three models such as (a) descriptive model, (b) predictive model and (c) prescriptive model (optimized). Sustainability of big data needs an intelligent model wherein cognitive intelligence along with artificial intelligence play a major role [7]. Therefore, neurogenerative machine learning techniques can be developed for future shape of metaverse. Data access, velocity and volume are taken care of by the storage technology. Quality of data is maintained by the cleaning part. Data analysis and visualization extract the values from the big data and accordingly human activities are predictively guided. The large amount of information may introduce specific difficulties which can be resolved by the big data technology. Thus big data has become essential to the victorious application of the virtual world. That is why basic thrust of this article is to present role of cognitive intelligence and artificial intelligence to sustain big data for creation of future metaverse. There have been several research works connecting metaverse with big data. Ooi et al. [4] explained the theories to express from the data outlook, the connectivity betweenthe real and virtual world. Cai et al. [8] illustrated the mathematics providing the solution to the computer and data intensive network in the metaverse. Smart phone data has been used to create metaverse by Park et al. [9]. A smart medical system is carried out by Yang et al. [10] which is formed on AI, Metaverse and data science.

We have organized this chapter in the following way: The fundamental concept of metaverse and the requirement of big data to ornament it are explained in Sect. 2. Sustainability of big data using cognitive intelligence, human-centric functional modelling and artificial intelligence is described in the Sect. 3. In the Sect. 4 different devices needed to conceptualize the future shape of metaverse is presented. Looking towards the societal benefits as an application of future metaverse, paradigm shift in education pattern (changes in teaching and learning atmosphere) is clearly exhibited in the Sect. 5. Since virtual reality, augmented reality, extended reality and mixed reality are backbone of future shape of metaverse, the Sect. 6 presents their role and also discusses the limitations of virtual reality in the context of establishment of future metaverse.

2 Fundamentals of Big Data and Metaverse

Big data is defined as that dataset which possesses large volume, velocity, variety, veracity and value. Big data addresses the cost effective information for decision making. Handling of big data pertaining to various industries is essential to develop corresponding models which are mainly descriptive, predictive and prescriptive. Predictive models provide forecasting of data, especially in the domain of financial sector, climate change and geology. Prescriptive models are built using optimization techniques and these models can be implemented to execute decision making problem, risk management and risk mitigation problems. Characteristics of big data can be further written as: (1) Volume: it refers to the amount of data. Data size is being increased from bytes to Yottabytes. The sources of big data can be obtained either from internet or from sensors or from field survey, (2) Velocity: it refers to the speed of data capturing and data processing, (3) Variety: this refers to data of structured, unstructured and semi-structured variety, (4) Veracity: this refers to biases, noise and uncertainty of data and (5) Value: this refers to the value that big data can provide, and it relates directly to what organizations can do with that collected data. A schematic diagram of big data is a shown in Fig. 2.

Big data plays an important role to create metaverse. Basically, metaverse is defined as an online environment in which users use their 3D digital avatars to engage in virtual activities similar to real-world activities such as developing lands, owning digital assets, and even utilizing digital cash that can be converted into virtual or real transactions. In short, metaverse is basically virtual reality enabled 3D internet which is our future internet. It is obvious that concept of virtual reality (VR) and augmented reality (AR) is required to know to build metaverse because the technology behind metaverse is the combination of VR and AR that develops extended reality (XR), platform on which metaverse stands. Applications of metaverseare based on different layers which are shown in Fig. 3. A large amount of data is generated in every layer on which the connectivity between the layers depends. So the operation of big data is very important for the metaverse market [11]. Big data technology creates the junction between the real world and the virtual world. As both the worlds are integrating

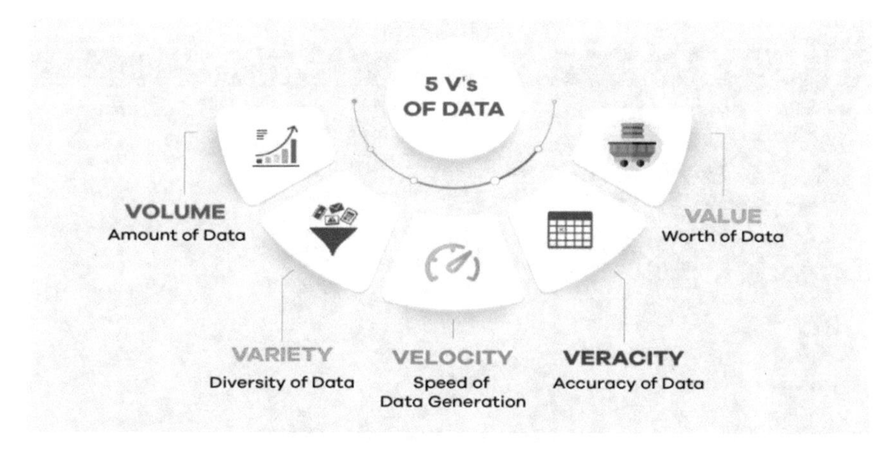

Fig. 2 '5 Vs' of Big Data

very fast, metaverse is emerging, too. Big data possess the following conditions to encounter the future metaverse. A typical market map of metaverse is as shown in Fig. 4.

Communication: Communication and storage can be combined with data using the technology of 5G and 6G [12, 13]. Related hardware and software that will surely be recommended by the metaverse era are big data, VR, AR, AI, cloud computing, block chain, cyber security etc.

Storage: Data will be stored in a decentralised process for unchallenging expansion, trouble free execution, standardized storage [14] and multi-copy consistency. Some applications need huge storage like AR, VR, streaming. This requirement will increase significantly in future digital world. Therefore the role of big data in metaverse is clearly visible.

Computational power: The computational power is very important as the conventional computation techniques are insufficient to fulfil the huge demand of the future.

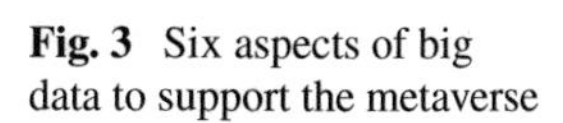

Fig. 3 Six aspects of big data to support the metaverse

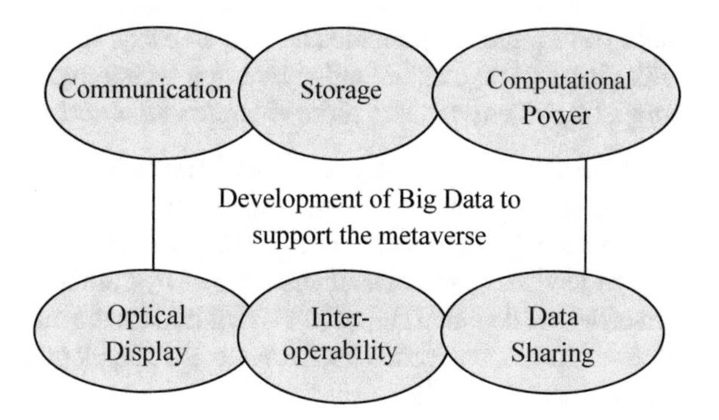

Fig. 4 Market map of the metaverse [11]

Thus metaverse needs computational power urgently where big data ensure the guarantee. The evolution of big data computing power will increase data interaction with which the creation of the metaverse will be accelerated [15].

Interoperability of data: In the metaverse the user creates their digital avatars to create their identity. These avatars can move across different platforms. This experience can be achieved uninterruptedly by data interoperability [16]. In other words in the metaverse an entity will collect the data and the data will move across multiple platforms. In metaverse different sub-metaverses exist. Data should be able to share information across the sub-metaverses to provide the users seamless experience [17].

Optical Display: The presentation of contents in the metaverse is important. For this purpose AR/VR devices, ultra HD, high speed accuracy of data transmission play important role. AR/VR need image processing and display. With the help of cloud storage and cloud computing, the development of cloud VR and AR can be promoted. This will greatly reduce the requirements for terminal equipment and make the equipment portable and easy to carry. In addition, cloud servers equipped with cloud storage and cloud computing technologies can make dataintensive and computing-intensive tasks more efficient and orderly [15].

Data sharing: Useful information can be provided to the service providers by sharing the data in the metaverse. Marketing and advertising people analyse the behavioural data to save the operational cost. Marketing developers improve the product based on user feedback. The consumers store a large amount of sensitive data in the metaverse. Secure data sharing system is required for the information exchange. Therefore blockchain based decentralized framework of data management is more appropriate for the metaverse [18].

3 Sustainability of Big Data

All over the world every day millions of digital data are generated for its post use by various scientific, industries and administrative companies. This large volume of data as collected, processed and utilized to discover the knowledge behind it is called Big Data. It is evidenced that big data is undoubtedly an important source of information for the private sector. Big Data plays a vital role in sustainable growth from the societal and institutional sphere. Basically, we need to transform big data into sustainable form. Now the question is which technology can be implemented for sustainability of big data?

To answer this question, let us first define Sustainability. Sustainability can be defined as "meeting the needs of the present without compromising the ability of future generations to meet their own needs." Sustainability is based on four features namely social, environmental, economic and cultural. The social aspect ensures the requirement to serve ourselves and others with respect. Environmental sustainability is to save the environment to maintain the nature and lives on the earth. Economic sustainability maintains the requirement to use the in hand resources in its best possible way for the value addition to human lives. Diversity and values are shared and nurtured by the cultural aspects.

Sustainable development by big data is possible due to its generation or collection, verification and correlation with information on social elements (light intensity per household, telephone calls, social networks activity, use of transport, etc.). Nowadays, there are several examples of the use of Big Data applied to sustainable development: The range and diversity of heterogeneous data is expressed in terms of variety. Velocity implies the speed with which the data are being gathered. Veracity portrays the truth or conformity of the data [19, 20]. The variability can be considered explaining the continuous changing characteristics of the data. With the technological growth the efficiency of creating and sharing information has significantly enhanced from 0.3 exabytes in 1986 to 65 exabytes in 2007 [21]. Almost everything depending on technology is now influenced by the bigdata analytics. Sustainability issues are not exceptions.

The objective of the application of big data to the environment is to achieve a better world for all. It has already become a powerful drive for observing and managing sustainable development. Big data can also help to ensure environmental and economic viability in the supply chain by establishing most efficient transit between raw material allocation, production areas, warehouses and customers. In the same way, the movement path of forklifts and other vehicles can be timely updated in the most efficient way through the real time monitoring of stock despatches thereby minimising fuel and energy consumption [22].

3.1 Sustainability of Big Data Using Cognitive Intelligence

Cognitive intelligence is nothing but human mental capability and understanding through thinking and experiences. Knowledge generation using existing information is the main physics of cognitive intelligence. It also includes other cognitive activities such as awareness, learning, recollection, judgment and reasoning. It can be said that cognitive intelligence is the ability of the human brain to digest information and form intelligence and meaning. Hence, measuring cognitive intelligence is crucial for organizations or industry handling big data for various types of jobs including recruitment and performance evaluation in the domain of teaching and learning, as it determines whether an applicant has the proneness to perform well at work that needs significant intellectual ability. It is said that cognitive intelligence uses practical knowledge that already exists and grows with implementation and experiences. Cognitive Computing facilitates detailed guidance towards building a new class of systems that learn from experience and derive insights to unlock the value of big data. Cognitive computing is the creation of self-learning systems that use data mining, pattern recognition and natural language processing to mirror the way the human brain works. The purpose of cognitive computing is to create computing systems that can solve complicated problems without constant human oversight. Today's cognitive computing solutions build on established concepts from artificial intelligence, natural language processing, ontologies, and leverage advances in big data management and analytics. Cognitive systems are rightly being hailed as the new era of computing.

Cognitive reasoning is the ability to analyse and perceive any given information from different perspectives by breaking it down into manageable components and structuring the information in a logical order. Therefore, if information is bundled in the form of a big data, cognitive reasoning will allow us to identify the partitioning of big data with respect to the problem in hand. Cognitive reasoning can be defined as an integral part of cognitive ability and is linked with an individual's ability to analyse facts rapidly, with transparency of thought and the ability to disregard unimportant information. Thus, to measure this reasoning, organizations should conduct cognitive reasoning tests. This is to infer whether a person can use logical steps to understand given information to assess what could be true or must be valid from the given data and directives. Cognitive function is defined as multiple brain-based activities that enable mankind to perform any given task. It involves the use of knowledge, information and reasoning to connect and understand information and derive meaning out of it.

Analysis of data by humans can be a time-consuming activity and thus use of advanced cognitive systems can be utilized to crunch this enormous amount of data [23]. As per Brian Krzanich (2016) "AI is based on the ability of machines to sense, reason, act and adapt based on learned experience" [24]. AI based system works on the rules and parameters that are fed inside it whereas a cognitive computing based system works by intercepting the command and then drawing inferences and suggesting possible solutions. In fact, cognitive computing is an AI based system that enables it to interact with humans like a fellow human, interpret the contextual meaning, analyse the past record of the user and draw deductions based on that interactive session. Cognitive computing helps humans in making decisions whereas AI based systems works on the concept that machines are capable of making better decisions on behalf of human. Cognitive computing is a sub-set of AI and anything that is cognitive is also AI. Data are captured from different sources, such as, mobile phones, web browsers and is stored in different data arrangements. These structured as well as unstructured data and their origins are difficult to be managed by the standard, conventional analysis and storage. A cost-effective decent software for big data analysis named Hadoop provides scalability, flexibility, accessibility, parallel processing as well as authentication. With the rise of the rate of development of data, machine learning methods are used for the automation of analysis of data like sentimental analysis (SA). The two important Vs that is velocity and volume of the data are huge, so the gap between the analysed data and the unprocessed data increases. This gap can be reduced by the advantage of a Big Data analytic solution. There are two Big Data processing outlook in SA: (i) batch processing, which is static and extremely controlled structure, that uses a distributed file system to store the data and after that a distributed computation structure is utilized. (ii) Interactive streaming data processing. For example, Apache Spark collects data as "streams" and the streaming data is processes through in-memory computational processes like Resilient Distributed Datasets (RDDs) [25]. Both Supervised and Unsupervised methods are used for SA which takes out essential data from the text data for helping the decision-makers. The disparity of the sentiment is effectively defined by Supervised method but for this purpose a huge number of labelled data are needed which is very difficult to achieve. On the other hand unsupervised methods, not being superior, can process data which are not so called labelled. Methods based on the support vector machine (SVM), Naïve Bayes (NB) and decision tree (DT) displayed better performance in SA. Like the classification problems, appropriate feature set selection is very important in this case. Some of the features extensively used in SA are word2vec, Parts of speech, Term Frequency (TF), Term Frequency Inverse Document Frequency (TF-IDF) etc. The use of a particular feature with a particular classification model shows particular outcome. Even so, a true classifier is not suitable for tweets and an integration ensemble or voting of many classifiers reveals superb efficiency for SA [26].

3.2 Cognitive Intelligence in 'Human-Centric Functional Modelling'

In the subsection, we have presented the role of cognitive intelligence for sustainability of big data which is basic ingredients for conceptualizing a virtual world to represent the physical world and this specific thought will give the shape of future metaverse. Therefore it is mandatory to discuss the knowhow of cognitive intelligence shaping humanoid (3D platform of various actions of humans as functions).

In other words we can say that conceptual spaces like Hilbert space, conventional Euclidian space, Sobolov Space, etc., can be formulated by functional state spaces where the main focus will be to develop the necessary mapping from one state to another state, and here the state will be represented as action of any human through cognition. So we can propose at this stage that future shape of metaverse will be based on "Human-Centric functional models (HCFM)" which will be formulated using cognitive intelligence. In this context, the concept of "state space" has been applied as a semantic approach by Bas C. van Fraassen, the philosopher [27] and later on, it is applied to all general sciences even for modelling a physical world. A state space is defined as 'the set of all possible configurations of a system' [28] whereas HCFM [29, 30] is defined as a model of human perception that explains the system to have a set of observable functions φ_i via which the change in functional state from "p" to "q" i.e. $\varphi_i(p) = q$ [31] will take place. The physical universe when visualized into a virtual universe using the mapping of states through various actions (functional) rather human cognition, we can say that we have achieved metaverse or we have created metaverse and that is our future shape of metaverse. Therefore the metaverse can be taken as a functional state space traversed by the consciousness and cognition of the universe. Graph is basically a network and using this nomenclature of network concept, the cognitive functional state space is explained by a directed graph with nodes (represent functional states) and the nodes are joined by edges (represent the reasoning processes) via which the transition takes place from one state to another. HCFM uses this graph network to express every possible concept connected by every possible reasoning process. A complete semantic model can be formulated provided the diagram of the functional state space of the cognitive system (the conceptual space) represents the complete human meaning of each concept and each reasoning process. If a conceptual space is to be represented as a "complete" semantic model, a number of assumptions come into the representation. One of the assumptions is that the meanings are to be self-contained in the region of the space and they should be carried through nodes and their links as shown in the Fig. 5.

With respect to the conceptual structure of metaverse can be as a combinatorial model of the real and a virtual world, the details of the world can be determined by the computation power that further calculates the speed of interaction between the worlds. This new concept of future metaverse shaping can be addressed as entanglement between physical and virtual world and hence it can be stated that the future of the metaverse will be related to quantum computing. But the speed of accessibility of stored data brings the limitations. So conventional quantum computing is not really appropriate for certain problems, particularly for which a large volume of input data is needed. The representation of metaverse as a functional state space, using HCFM may radically accelerate research in quantum computing to be explored.

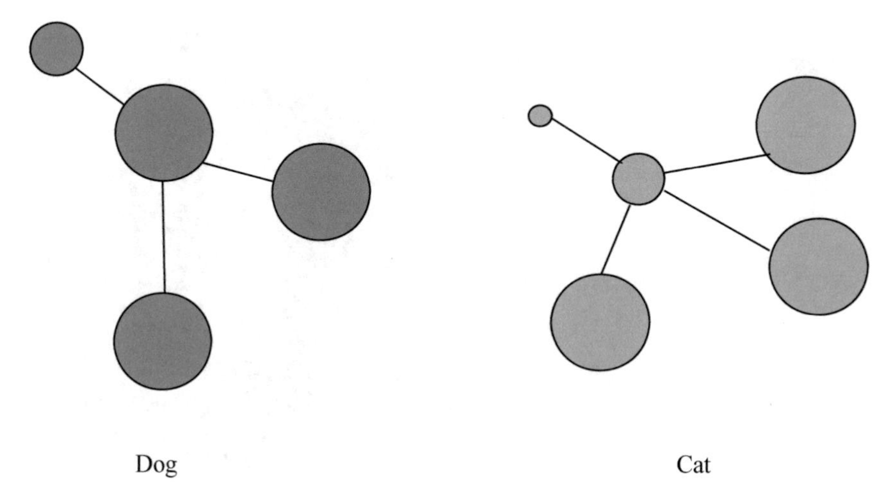

Fig. 5 Conceptual depiction of meaning

3.3 Artificial Intelligence for Sustainability of Big Data

Artificial intelligence plays a big role for sustainability big data. AI based metaverse is defined as multiple hybrid-platform worlds (a shared digital 3D world) that allows users to experience a comprehensive immersive sensation with interactive and synergic activities. In this case various types of objects, user identities, and digital commodities can be exchanged between virtual worlds and even reflected into the real world. A metaverse platform consists of many layers (see Fig. 6) which are (a) Infrastructure (5G, 6G, WiFi, cloud, data center, central processing units, and GPUs), (b) Human interface (mobile, smart watch, smart glasses, wearable devices, head-mounted display, gestures, voice and electrode bundle), (c) Decentralization (edge computing, AI agents, blockchain, and microservices), (d) Spatial computing (3D engines, VR, augmented reality (AR), XR, geospatial mapping, and multitasking), (e) Creator economy (design tools, asset markets, Ecommerce, and workflow), (f) Discovery (advertising networks, virtual stores, social curation, ratings, avatar, and chatbot), (g) Experience (games, social, E-sports, shopping, festivals, events, learning, and working).

Future shape of metaverse will be completely based on CI and AI. Progress of future metaverse will be driven by AI powered blockchain, XR/VR, and 5G and guided by CI for cognition. In this context, quantum computing will ease the speed of computation due to superposition principle of qubits structure and entanglement. However, there will be an uncertainty regarding the influence and effective contribution of AI in shaping metaverse. The reason behind this is it is neither introduced in a creative way like XR/VR nor explained attractively on social media as blockchain. No existing work provides a complete review of the role and application of AI in the metaverse.

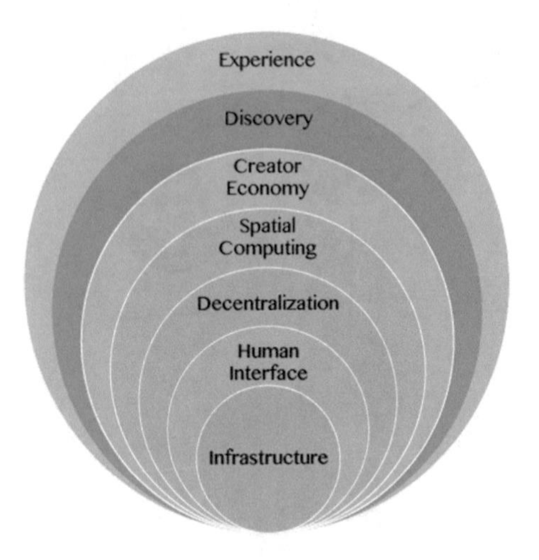

Fig. 6 Seven layers of a metaverse platform

If AI is combined with technologies like AR/VR, networking and block chain, it is possible to create an assured, scalable, and pragmatic digital worlds on a reliable platform in the metaverse. The importance of AI is undoubtedly very clear in improving the performance and to assure the reliability, if the seven-layer metaverse platform is taken into consideration. Many modern machine learning algorithms in the 5G and upcoming 6G systems, can be merged with reinforcement and supervised learning to perform different rigorous tasks like monitoring efficient spectrum, allocation of automatic assets, to prevent attack, to detect network fault, estimation of channel, traffic off-loading etc. Analysis of both simple and complicated human activities can be possible using wearable gadgets based on sensor mechanism and other man–machine interaction gadgets based on machine learning and deep learning models. Therefore the real world activities of users' are extrapolated into the virtual worlds and the users are allowed to control their digital identities (avatars) for interaction in the metaverse smoothly. These avatars recreate many emotional expressions adopted from the real world, such as facial impressions, movement of body, other physical interactions, other than speech recognition of speech and thought analysis, which are controlled by AI in terms of accuracy and processing speed.

It is true that VR/AR/XR technologies empower the user to enter in the metaverse, with immersive gadgets, such as glasses, head-mounted displays, gloves, but AI is a vital technology which is responsible to create the virtual world beautiful. This also generates a flawless and smooth virtual-reality sensation to users. The process of content creation is also facilitated by AI, for example, NVIDIA has introduced GANverse3D, an AI module, which enable users to take photographs of objects and then virtual replicas are made. Many procedures based on deep learning have been suggested for building 3D objects which includes human body parts also. Software (e.g., PyTorch3Dlibrary from Facebook AI and Tensor RT from NVIDIA) as well as hardware (e.g., GPUs) both are available for excellent accuracy while presenting real time processing. Very recently the AI research super cluster (RSC) has been introduced by Meta. It is believed that it is one of the fastest AI supercomputers to accelerate AI research and may be served for creating the metaverse. Furthermore, better deep learning models can be developed with the help of RSC from the huge amount of heterogeneous data for different applications. As a result, RSC driven outcomes and successes are to be used to create the metaverse platform, in which AI will play an important role.

4 Metaverse Devices

Metaverse is a virtual domain to be explored which will offer people to be engaged with activities which are not possible in the real world. In future possibilities are huge and some extreme creative ideas are on the table already. Even people are trying to sell virtual land in the metaverse. Big companies are already interested in it. Businesses are ready to accept its potentialities though it is in a budding stage. Some VR gadgets are also ready to enter within the virtual realm. One can use a variety of devices with an input system, a screen and an appropriate processor. It is true that people can access metaverse through smart phones and computers but not in the fully hypnotic way. For the mesmerizing sensation specific metaverse devices are needed to be immersed fully in the world as they provide something special that adapt people's expectations. VR eye glasses or AR gear actually feeds the metaverse into the user's field of vision. Then the user can interact with the digital domain using the specific input devices. Some of the metaverse gadgets are (Fig. 7).

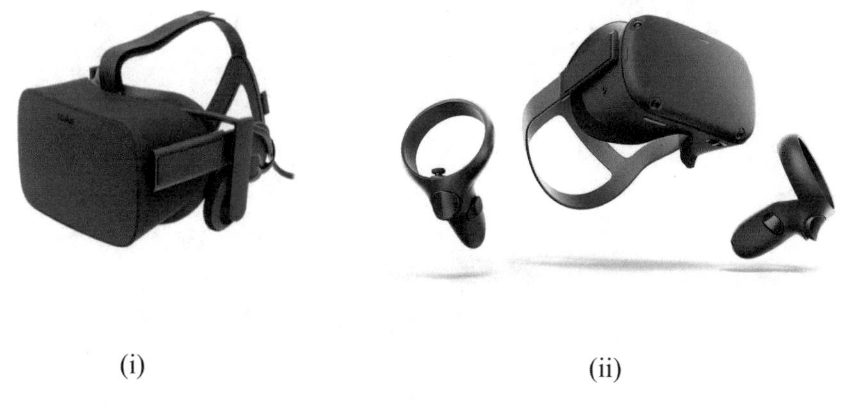

(i) (ii)

Fig. 7 (i) Oculus Rift CV1 VR Headset, (ii) Oculus Quest

4.1 VR Metaverse Gadgets

1. Oculus Rift & Oculus Quest

Oculus Rift is a VR headset which is planned to work with a PC for processing. On the other hand Oculus Quest 2 is a headset that works on the android system which is rebranded as Meta Quest 2 (Fig. 8).

2. Valve Index

Valve Index is a second generation VR headset manufactured by Valve. The headset is designed in such a way that skin contact is reduced. A fabric lining inside prevents sweat and thus infections from microorganisms. The Valve Index controllersconsist of a touchpad, a joystick, a menu button, two face buttons and a trigger. Controllers are allowed to track hand and finger position, movement by 87 in-built sensors and a perfect representation of the use's hand is created in the virtual reality. The Valve Index is wired device with a head mounted display. It needs external stations for tracking and wired computer interface thus it not so called 'user friendly'. But the graphics quality it provides is excellent (Fig. 9).

3. Play Station VR2

Play Station VR2 (PSVR2) is an approaching VR head set for the PlayStation 5 (PS5) home video game console (Figs. 10 and 11).

VR2 uses the cable connection that connects with PS5. For the effective immersion and sensitivity eye-tracking is used in PSVR2 which is supported by 3D audio and sensory function for high end in-game activities. PlayStation 5 has been announced in 2019 as next in line to PlayStation 4. It was released on 12.11.2020. In August 2021, a hardware revision of PS5 has been done.

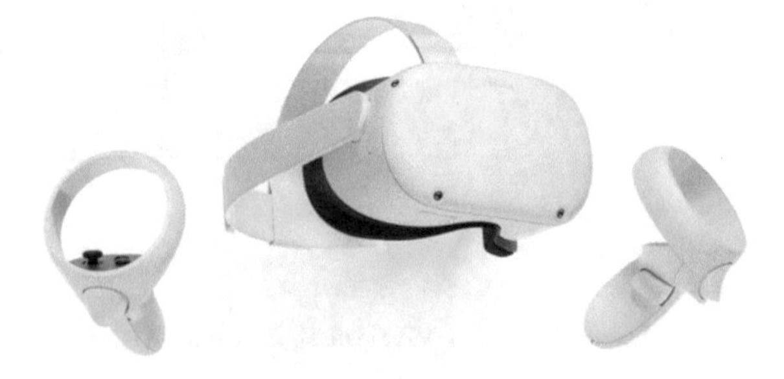

Fig. 8 Meta Quest 2 Headset

Fig. 9 Valve Index VR Kit

Fig. 10 PlayStation VR2

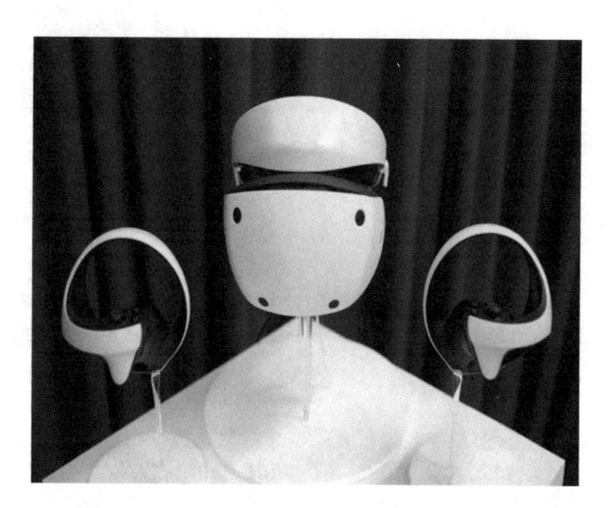

4.2 AR Metaverse Gadgets

1. HoloLens

Micrisoft'sHoloLens is one of their most important ventures for augmented reality. It handles lots of optical sensors, microphones, cameras. Another significant feature "Holographic Processing Unit" is present in HoloLens making it one of the best metaverse devices for Augmented Reality. This is a premature project, hand motions are yet to improve. Microsoft Hololens 2 are the next generation devices for working

Fig. 11 PlayStation 5

Fig. 12 HoloLens by Microsoft

with AR/VR (mixed reality) contents with an increased angle of view and more immersion. The users are allowed to interact with holograms (Figs. 12 and 13).

2. Epson Moverio

The Epson Moverio is AR smart glasses with Si-OLDED display with extraordinary vision clarity. It offers 34° wide field of vision, fully HD, high contrast and high resolution. The Moverio BT-300 weigh 20% lighter compared to its predecessor, making it the lightest AR headset in the world. It works on the android system. This is the advantage as well as disadvantage at the same time of the gadget. The disadvantage is that the system must adopt the software, and the disadvantage is that the application of android is not that complicated.

3. Haptic Gloves and bodysuit

Haptic gloves will launch a tactile experience to the virtual realm. It contains a series of soft motors motors, actuators, sensors and other tech wizardry which will allow users to grip and touch virtual objects while interaction in the metaverse. Meta-owned Reality Labs are working to reproduce a range of real-world sensations like

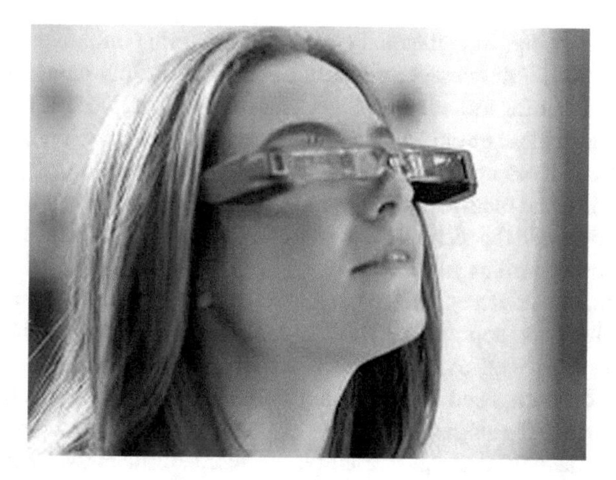

Fig. 13 Moverio-BT-300

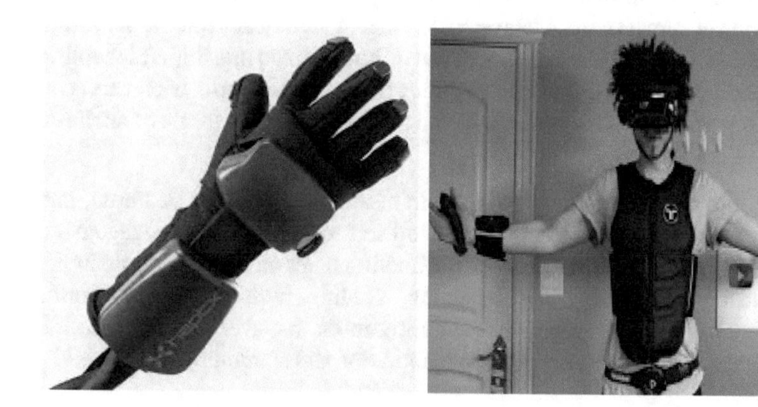

Fig. 14 Haptic Gloves and bodysuit

texture, pressure and vibration. Haptic bodysuits capture motions through electro-simulation sensors. When teamed with a VR headset it will dramatically improve the interaction in the virtual world by adding sense of physical touch to our digital experience (Fig. 14).

5 Change of Teaching Methodology in the Metaverse

From the beginning of 2020, the education system has been thrown into a disorder due to the COVID-19 pandemic. It is declared by the UNESCO that it "constitutes the worst education crisis on record". As a result, many education systems introduced online remote learning. But "the quality of remote learning initiatives varied greatly, as did access" [23]. On average, globally schools were fully closed for 20 weeks and partially closed for more than 20 weeks. Students lost two trillions of teaching hours [24]. Under these circumstances virtual learning was given great importance. But the ineffectiveness and discrepancies in the quality of learning, problems with accessibility to virtual platforms were there of course. So the need for better teaching–learning experience in the virtual world was created globally to reduce the disparities in the education system as well.

As the demand and acceptance of online learning increases the whole education system is looking for appropriate ways to make the teaching learning experience more interacting as well as effective. It can be achieved by keeping the current generation learners engaged which is very difficult. And here comes the metaverse giving the chance to the learners and the teachers to learn and interact in a creative and focussed way to achieve the teaching learning outcomes efficiently. Here students can touch, grab, walk around the learning objects, can be immersed into activities in the laboratories, historical events, even in the space with lots of more opportunities and experiences which are impossible to achieve in the physical world. This will improve the retention power remarkably. Metaverse can take a significant role in the future shape of education. Conventional schooling can be transformed into digital schooling with the immense possibilities and opportunities in the digital world. Metaverse can satisfy the best principles for learning which can be expressed in terms of six basic criteria- 6 Cs of learning.

1. Collaboration: Social interactions promote aesthetic perception, learning, and community formation which help to develop self regulation while engagement. In children's growing years elementary schooling takes an important role in it.
2. Communication: speaking, understanding, reading, writing together forms communication. Though in-person mentoring in the metaverse is difficult, but with the advancement of AR, VR, XR collaboration and communication is child's play now.
3. Content: In conventional teaching learning method usually paper contents are used which is difficult for customised experiences. In the metaverse with endless technical possibilities customised contents can be offered to the students according to their need to achieve the best outcome. In this process collaboration and communication are involved any ways.
4. Critical Thinking: Critical thinking is a necessary skill for decision making. It is important to make the children to be able to reason. They must learn to communicate and collaborate among themselves to enhance their point of views and express their own opinion logically. In the metaverse this will happen.
5. Creative Innovation: Children can create innovation by combining the critical thinking with unlimited contents offered. When the contents are converted to games in the virtual world it provides the audio-visual aids as well as let them visualise what they are learning. Play-oriented teaching learning makes the students more curious and encourage to explore the subject.

6. Confidence: If a learner is capable to use the 5 Cs comfortably, the 6th will come naturally.

Technology can overcome almost all obstacles and it will do the same in the domain of education (Figs. 15 and 16). As an example of learning experience in the virtual world the involved technologies can be compared as follows [25] (Table 1).

The technology of Eduverse has been proven helpful for many students in the classroom trough 3D animation, storytelling etc. Metaverse will represent the upcoming stage of the evolution which will make the students feel the real world situation virtually and make learning fun. Educators must offer interactive curriculum instead of conventional or traditional one that can take the opportunities of real time experience using digital technologies. In spite of the fact that metaverse makes the learning easy for students by engagement, efficacy should always be measured by the results it shows. Student's grades and achievements can only convince the teachers, the parents and otherteaching fraternities. It was already proved througha survey that

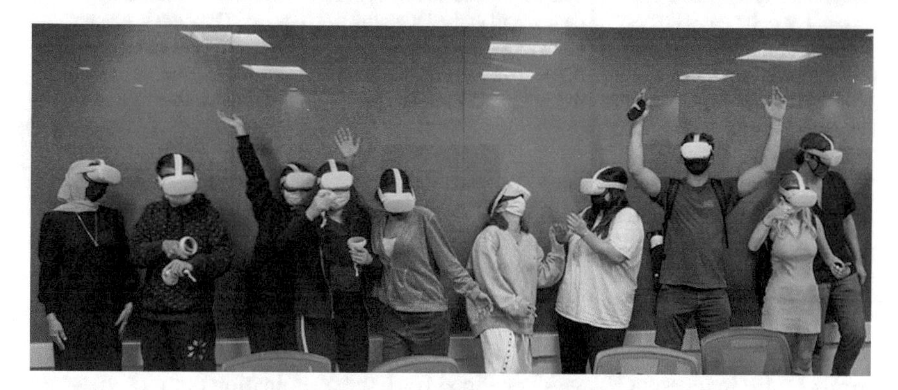

Fig. 15 Learning in the Metaverse-UM school of Communication

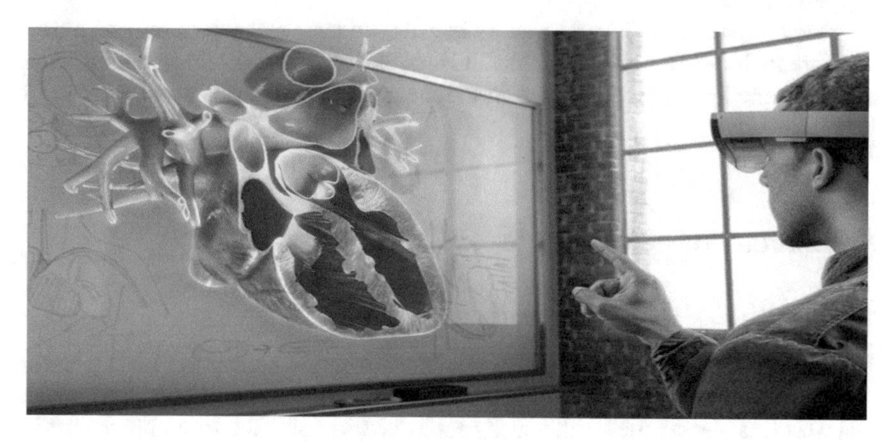

Fig. 16 Student in Life Science Class

Table 1 Virtuality in learning methods

Augmented Reality (AR)	Mixed Reality (MR)	Virtual Reality (VR)
Gadget: From the uses' view digital contents are superimposed on the real platform using mobile device or see through head-mounted display	Gadget: Digital contents are merged with the physical world of the users' using see-through head-mounted display	Gadget: Using head-mounted display users can enter and interact with a technology generated virtual world
Description: Interactions between real and virtual objects face several limitations. For example, 3D objects may be semi-transparent and cannot hide behind physical objects	Description: More natural interactions between physical and digital objects. For example, virtual characters can be concealed behind real objects	Description: Audio-visual stimuli from the real world are replaced with those of the computer generated world
Scenario: Two students in the same place scan an image of a fish using smartphones. Then an animated 3D model of the species is formed overlaying text information. A person in an AR head-set display can see the same 3D model from their points of view	Scenario: Two students in different places scan their unique physical space by a MR device and embed a digital coral reef onto it where a digital fish can swim around the place and each learner	Scenario: Two students in different places are on a virtual coral reef using a VR device They are scuba divers (avatar) in the virtual space and can interact with the fish and one another in that space
Hardware: Smartphone, tablet, AR head-mounted display like Hololens, Realear devices, Magic Leap 1, or ThinkReality A3	Hardware: Hololens, Magic Leap 1, Varjo XR3, Lynx	Hardware: Meta Quest 2, HTC Vive

learning with immersive VR was beneficial from the examination point of view. VR learning has been established to be an unconventional substitute for conventional education system.

It very important to follow carefully how the students understand the metaverse, what are their needs in the metaverse. It is necessary to analyze the pattern of their activities, the level of engagement in the metaverse and the outcome as a part of learning.

6 Limitations of Metaverse

There are darker sides, too. The limitations are:

Health issues: VR "Hangovers" are common. Post VR depression is dangerous. After experiencing an amazingly immersive world when the user comes back to the real world, sadness comes with which people have to fight. Too much immersion in the digital world and isolating ourselves from the practical world can increase the chance of psychosis.

Addiction: As the user is completely immersed in the virtual world, there is a huge risk of addiction in the metaverse which can influence the development of the kids and teenagers. Users who are immersed in the metaverse for a long time, might be detached to reality and refuse to accept the existence of a real world besides the virtual world.

Privacy issues: With digitalization come privacy challenges. Our online activities are being traced when we browse the web. This tracing is going to be more enormous in the metaverse. For example, eye-tracking technology is used in VR-Headsets, so from the headset data advertisers can always track us. Our physical reactions are going to be monitored through the wearable devices that measure our feelings and sensations.

Metaverse Laws: There is a chance that metaverse is going to be a lawless territory as virtual activities cannot be crimes. This makes it extremely unprotected and vulnerable to different sorts of illegal ventures like fraud, child exploitation, cyberattacks etc. It can be a significant risk for young generation with less experience.

Desensitization: While playing violent games in VR the users can touch and feel what they are doing. This can lead to people becoming desensitized to their activities like shooting a gun at someone. As a result brutality, aggression, inequality and racism come. If people start imitating their online activities offline, that will be a serious threat.

Identity issues: It is not that difficult for people to hack our avatars, destroy our online identity in the metaverse. So focus on security in the metaverse is needed.

Socialization issues: People immersed in the virtual world for a long time do not want to be connected to their own society rather they like to create a new isolated society. As a result their cultural behavior becomes restricted and affected which eventually destroy many traditions that have existed from the beginning of mankind.

References

1. Stephenson, N.: Snow crash: A novel. Spectra (2003)
2. Thien, H.-T., Pham, Q.-V., Pham, X.-Q., Nguyen, T.T., Han, Z., Kim, D.-S.: Artificial Intelligence for the Metaverse: A Survey, arXiv:2202.10336v1, 15 Feb (2002)
3. Radoff, J.: The metaverse value-chain. Available: https://medium.com/building-the-metaverse/the-metaverse-value-chain-afcf9e09e3a7, Apr. 2021
4. Ooi, B.C., Tan, K.-L., Tung, A., Chen, G., Shou, M.Z., Xiao, X., Zhang, M.: Sense the physical, walkthrough the virtual, manage the metaverse: a data-centric perspective, arXiv preprint arXiv: 2206.10326 (2022)
5. Mohammadi, M., Al-Fuqaha, A., Sorour, S., Guizani, M.: Deep learning for IoT big data and streaming analytics: a survey. IEEE Commun. Surv. Tutorials **20**(4), 2923–2960 (2018)

6. Sagiroglu, S., Sinanc, D.: Big data: a review. In: International Conference on Collaboration Technologies and Systems, pp. 42–47, IEEE (2013)
7. Ge, M., Bangui, H., Buhnova, B.: Big data for internet of things: a survey. Futur. Gener. Comput. Syst. **87**, 601–614 (2018)
8. Cai, Y., Llorca, J., Tulino, A.M., Molisch, A.F.: Computeand data-intensive networks: the key to the Metaverse, arXiv preprint arXiv:2204.02001 (2022)
9. Park, D., Kim, J.M., Jung, J., Choi, S.: Method to create a metaverse using smartphone data. In: International Conference on Human-Computer Interaction, pp. 45–57. Springer (2022)
10. Yang, Y., Siau, K., Xie, W., Sun, Y.: Smart health intelligent healthcare systems in the metaverse, artificial intelligence, and data science era. J. Organ. End User Comput. **34**(1), 1–14 (2022)
11. Lee, L.-H., Braud, T., Zhou, P., Wang, L., Xu, D., Lin, Z., Kumar, A., Bermejo, C., Hui, P.: All one needs to know about metaverse: a complete survey on technological singularity, virtual ecosystem, and research agenda, arXiv preprint arXiv:2110.05352 (2021)
12. Ning, H., Wang, H., Lin, Y., Wang, W., Dhelim, S., Farha, F., Ding, J., Daneshmand, M.: A survey on metaverse: the state-of-theart, technologies, applications, and challenges, arXiv preprint arXiv:2111.09673 (2021)
13. Gupta, R., Kumari, A., Tanwar, S.: Fusion of blockchain and artificial intelligence for secure drone networking underlying 5g communications. Trans. Emerg. Telecommun. Technol. **32**(1), e4176 (2021)
14. Al-Kiswany, S., Gharaibeh, A., Ripeanu, M.: The case for a versatile storage system. ACM SIGOPS Operat. Syst. Rev. **44**(1), 10–14 (2010)
15. Tang, X., Cao, C., Wang, Y., Zhang, S., Liu, Y., Li, M., He, T.: Computing power network: the architecture of convergence of computing and networking towards 6g requirement. China Commun. **18**(2), 175–185 (2021)
16. Lim, W.Y.B., Xiong, Z., Niyato, D., Cao, X., Miao, C., Sun, S., Yang, Q.: Realizing the metaverse with edge intelligence: a match made in heaven, arXiv preprint arXiv:2201.01634 (2022)
17. Nawaratne, R., Alahakoon, D., De Silva, D., Chhetri, P., Chilamkurti, N.: Self-evolving intelligent algorithms for facilitating data interoperability in IoT environments. Futur. Gener. Comput. Syst. **86**, 421–432 (2018)
18. Chen, Y., Guo, J., Li, C., Ren, W.: FaDe: a blockchain-based fair data exchange scheme for big data sharing. Future Internet **11**(11), 225 (2019)
19. Russom, P.: Big data analytics. Data Warehousing Inst. **4**(1), 1–36 (2011)
20. Abbasi, A., Sarker, S., Chiang, R.H.: Big data research in information systems: toward an inclusive research agenda. J. Assoc. Inf. Syst. **17**(2), 3 (2016)
21. Manyika, J.: Big data: The Next Frontier for Innovation, Competition, and Productivity. McKinsey Global Institute, Washington (2011)
22. Nancy Master: What is Big Data's Role in Sustainability and Supply Chain Innovation? (2017) [Online] Available at: http://www.rfgen.com/blog/what-is-big-datas-role-in-sustainability-and supply-chain-innovation. Accessed 29 December 2017
23. Kim, H.W., Chan, H.C., Gupta, S.: Examining information systems infusion from a user commitment perspective. Inf. Technol. People **29**(No. 1), 173–199 (2016)
24. Krzanich, B.: The intelligence revolution—Intel's AI commitments to deliver a better world. Web-LinkEditorial, Intel Newsroom https://newsroom.intel.com/editorials/krzanich-ai-day/ (2016)
25. Nair, L.R. et al.: Applying spark based machine learning model on streaming big data for health status prediction. Comput. Electr. Eng. (2018)
26. Rustam, F. et al.: Tweets classification on the base of sentiments for US airline companies. Entropy (2019)
27. Lloyd, E.A.: A semantic approach to the structure of population genetics. Philos. Sci. **51**(2), 242–264 (1984)
28. Nykamp, D.: State space definition. Math Insights. Retrieved 17 November 2019 (2019)
29. Williams, A.E.: (Accepted) A Revolution in Systems Thinking?, Proceedings 2021 Congress of the World Organization of Systems and Cybernetics (WOSC)

30. Williams, A.E.: Human-centric functional modeling and the unification of systems thinking approaches: a short communication. J. Syst. Thinking, 5–5 (2021)
31. Laubenbacher, R., Pareigis, B.: Equivalence relations on finite dynamical systems. Adv. Appl. Math. **26**(3), 237–251 (2001). https://doi.org/10.1006/aama.2000.0717

Towards 3D Virtual Dressing Room Based User-Friendly Metaverse Strategy

Mahmoud Y. Shams, Omar M. Elzeki, and Hanaa Salem Marie

Abstract This paper applies a meta-verse strategy as an intelligent application works as a virtual reality for the dressing room. The proposed website is designed as a virtual fashion store, such that the women and/or men may foumd their fashion with a privacy and convenience characteristics based on the E-commerce and meta-verse strategy. The proposed website is a real-time virtual dressing room that eliminates a lot of the trouble from shopping by which no more long time required in the fitting room with an armful of clothing or the time-consuming process of getting dressed and undressed multiple times. Stand a few feet in front of a webcam to utilize this device, which uses motion-sensing technology. On the computer screen, a live image of you will display, along with various categories such as trousers, shirts, and dresses. By waving your hand over Translation Controls, Scale Controls, and Selection Controls, you may select a category, such as shirts and the preferred colors. Therefore, the clothing you wish to try on will be on, and it will be digitally overlaid over your live image. A quick size algorithm to assist the user in determining the prices and appropriate size is proposed. Moreover, we present RSA for user authentication to make the system more secure especially for mobility clients. As a result, the primary rule is that the proposed meta-verse strategy should generate pictures in real-time and respond to user interaction in real-time. It should also be a low-cost, user-friendly system.

Mahmoud Y. Shams: Scientific Research Group in Egypt

M. Y. Shams (✉)
Faculty of Artificial Intelligence, Kafrelsheikh University, Kafr El-Sheikh 33516, Egypt
e-mail: mahmoud.yasin@ai.kfs.edu.eg

O. M. Elzeki
Faculty of Computers and Information Sciences, Mansoura University, Mansoura 35516, Egypt
e-mail: omar_m_elzeki@mans.edu.eg

H. S. Marie
Faculty of Engineering, Delta University for Science and Technology, Gamasa 35712, Egypt
e-mail: hana.salem@deltauniv.edu.eg

O. M. Elzeki
Faculty of Computer Science & Engineering, New Mansoura University, Gamasa 35712, Egypt

A. E. Hassanien et al. (eds.), *The Future of Metaverse in the Virtual Era and Physical World*, Studies in Big Data 123, https://doi.org/10.1007/978-3-031-29132-6_2

Keywords 3D dressing room · Real-time · Website · Meta-verse

1 Introduction

Over the past few decades, digital manufacturing has provided significant benefits to the whole sector. Digital manufacturing develops models and mimics product and process improvement by digitally simulating factories, resources, workforces, etc. [1]. The metaverse, according to analyst Forrester, is an immersive experience of interconnected and interoperable worlds that will be made available through a number of devices. A decentralised platform that overlays a 3D experience on top of the World Wide Web will be supplied gradually, according to the Forrester research "*The status of the metaverse*". According to Forrester, there are potential to build out meta-verse capabilities across all industry sectors, although adoption among consumers is most likely to come from gamers and those who are engaged on social media sites. According to preliminary Forrester research, 22% of online adults in the United States and the United Kingdom engage in intense gaming and social media activities that are favourable to metaverse early adoption [2].

Online massively multiplayer games are no longer just games. "Metaverses," a subset of these games that includes Second Life and Active Worlds, represent some of the most immersive, interactive learning, simulation, and digital design possibilities available today. They also blur the lines between work and play, as well as between user and designer, raising questions about the nature and practise of virtual design, or design practised inside virtual reality by and on 3D avatars [3].

In the daily lives, it will soon have a digital layer added within a Real-World Metaverse powered by real-time data, a virtual version of ourselves will carry out all of our regular tasks. Through the use of meta-verse, we shall spend our entire lives in virtual settings. With immersive experience and digital transformation, a variety of virtual environments have been created, including social networks and virtual gaming worlds, with thousands of services and apps. However, most of these environments lack coherence and are not connected into a platform [4, 5].

With the goal of automatic generation of direction-giving gestures in Metaverse avatars, an empirical study was conducted by Tsukamoto et al. [6] to collect human gestures in directiongiving dialogues. Then, they looked into the relationship between proxemics and gesture distribution. Therhy presented four different types of proxemics based on their distance from the gesture display space. Finally, they suggested a mechanism that uses proxemics information and language information from chat text to determine the timing and form of avatar gestures. Also, they demonstrated how the their mechanism can generate animation and speech on a Second Life Metaverse avatar.

Digital twinning is one of the latest concepts of modern technology, which integrates different sectors into a single technical system through which different information is exchanged [7]. The main driver behind digital twinning is artificial intelligence technologies. These technologies help some machines perform certain tasks without being programmed to do those tasks [8, 9].

There are now so many different online dressing rooms that Macy launched a new changing room in October 2010, enabling consumers to choose products from a computer using an iPad and a touch mirror [10]. Within secure internet access, a password is required for authentication [11]. The shopper can share the image of the merchandise with friends online and then enlarge the garment to fit the desired size. The forms of Virtual Dressing Room (VDR) generally consist of four categories investigated in Fig. 1. The first category is Video VDR (VVDR) that employs video technology to get a lively and realistic view of the clothes they like [12]. The second category is Robotics VDR (RVDR) [13] by which the first was invented by an Estonian startup called "Fits. Me" [14]. The robot prototype has been in development, and currently, it can transform into 2,000 body types. The "Fit Me" robot program requires various measurements to accurately show the right robot body style for the buyer. Once the results have been obtained, consumers can virtually try on different sizes of clothing to determine the best fit. The third category is the Motion Detector VDR (MDVDR), such that in hand motion, consumers can refer to different outfits on the screen to try them on [15]. A wave of your hand can add items to the virtual shopping cart, which counts them and adds them to the screen. With another wave, consumers can send virtual clothes pictures to their mobile phones.

The fourth category is the Webcam VDR (WVDR), by which the VDR depends on the webcam camera through a secure website [16]. Zugara introduced the first virtual dressing room on the Web camera in June 2009, as previously reported [17]. Furthermore, Tobi.com was the first online retailer that launched fashion products in November 2009. To use the Fashionista program, the online buyer prints the bookmark and then follows the online procedure to align the actual position with the webcam. The clothing is correctly aligned on the monitor [18]. The later version of the Social Software Webcam Camera eliminates the need for bookmarks. Instead, the

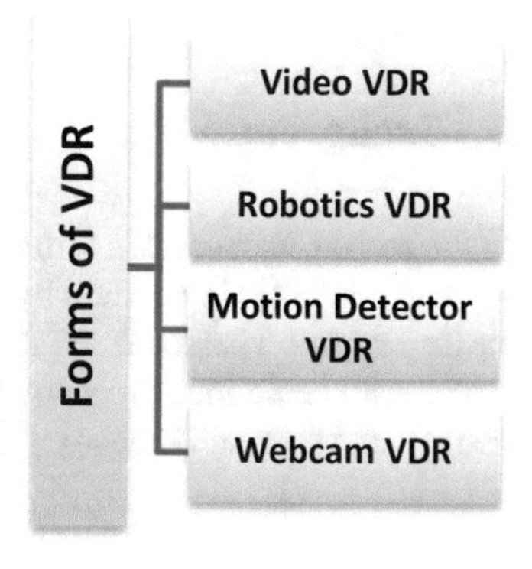

Fig. 1 The general forms of virtual dressing room based digital twining

problems of the face and aligns the destination positioning with the webcam. Other new features of the virtual fitting room include an interactive mirror that uses virtual technology to personalize the real-world shopping experience. Paxar, a subsidiary of Avery Dennison, launched RFID mirrors in 2007. Shoppers wearing RFID-tagged clothing standing in front of the RFID mirror will automatically be greeted by the mirror, which contains various information about the clothing, including materials and colors to choose from [19]. The mirror will also provide a selection of accessories and suggest different items combined for elegant coordination. Custom technology, such as the Intelligent system, has not replaced the virtual dressing room. In 2005, Levi' launched and tested the Intellifit system, which uses radio-wave technology to scan the contours of your body to customize a near-perfect fit. In this chapter, virtual dressing rooms are presented as one of the meta-vers application to deliver augmented reality technology that enables the user to put clothes on before purchasing. In addition, we employed virtual computer cameras to display online customers how the garments would look on their bodies in real-time (video), allowing them to interact with the interface using motion controls of 3D avatars. The Webcam Social Shopper is a mobile software that transforms online clothing buying into a social, engaging, and enjoyable experience.

The main objectives of this work include the importance of digital twinning to connect the real and virtual dressing rooms towards meta-verse environment strategies are listed as follow:

- After the website registration is completed, the website will provide you with a virtual dressing room in real-time.
- Eliminate the need for changing rooms in physical stores.
- It is not necessary to try on clothes to see if they fit.
- If the user chooses to sell a T-shirt, they can change the width of the shoulders, the length of the T-shirt, and the waist circumference of the T-shirt. Then when the user selects other shirts, the same measurement values will also be applied to these shirts based on these measurement values.
- Very user-friendly shopping task.
- Connection of augmented physical reality with virtual reality to met the meta-verse strategy requirements.

2 Background

In the meta-verse world, a virtual human can enhance the user's immersion through human–computer interaction by mirroring the user's motions and emotions. How to easily generate a personalised virtual human, as well as how to mirror a real human's motions and emotions to a virtual human, have become hot research topics in virtual reality academia and industry [20].

Generally, there are many attempts to utilize and create VDR. Isikdogan and Kara [12] present a VDR application based on Kinect Microsoft sensor by extracting the enrolled user from a video stream using labeled user data that register the tracking data of the clothes model. Virtual Reality (VR) and Augmented Reality (AR) are considered as the most important techniques used in VDR such that they reduce consumers' perceived risk of online clothing shopping and have a positive impact on consumer self-confidence, consumer-brand relationships, and consumer online shopping behavior [16].

A system based on a basic open library and Microsoft Kinect to assess the user experience on adaptive performance key metrics such as attention, significance, and alertness revealed that 96% of respondents were satisfied with the usage of VDR is provided in [15]. The relationships between shopping satisfaction, commercial inspiration, clothes shopping self-assurance, observed informativeness, perceived annoyance, and buying intention, using an intermediated model-based VDR are illustrated [21].

A single V2 sensor is used as a concentration sensor in Microsoft Kinect version two (V2) to record measurements of the user's body characteristics, including 3D measures such as breast, waist, hip, thigh, and knee, to construct a unique model for each user was presented in [22]. The clothing size list is selected based on the measurements of each customer. Six key challenges were derived from the interviews, coming from market-related, technology-based, and company-specific topics. In response to this scenario, six key success factors are outlined: vision and strategy, overall underlying processes, user interface and communication, custom technology solutions, supply chain engagement, and change management [23].

Bidirectional twinning is based on converting the physical environment into a virtual environment and vice versa, as shown in Fig. 2. While the transition from the physical environment to the virtual environment depends on the collection of hardware, the transition from the virtual environment to the physical environment requires embedded data extracted from the virtual environment. These data allow the operation applied to the physical environment [24].

In this work, the initialization process includes the following steps. The avatar creation step by which the user stands in front of the mirror with "P" pose to take images of their body using a webcam. To reflect the precise size of the user's, lightly user clothed is highly required. One Time the avatars are created; they can be utilized if the size of the customer's body does not alter. The second phase is known as user selection, and it involves the customer selecting garments from the Clothing Stores (CS) and Clothes Type (CT) windows to simulate with the avatar. The third phase is scaling, which determines the scale of the selected clothing based on the user's hand movements and interaction with the translation, scale, and selection buttons, as well as detection based on the adjusted webcam.

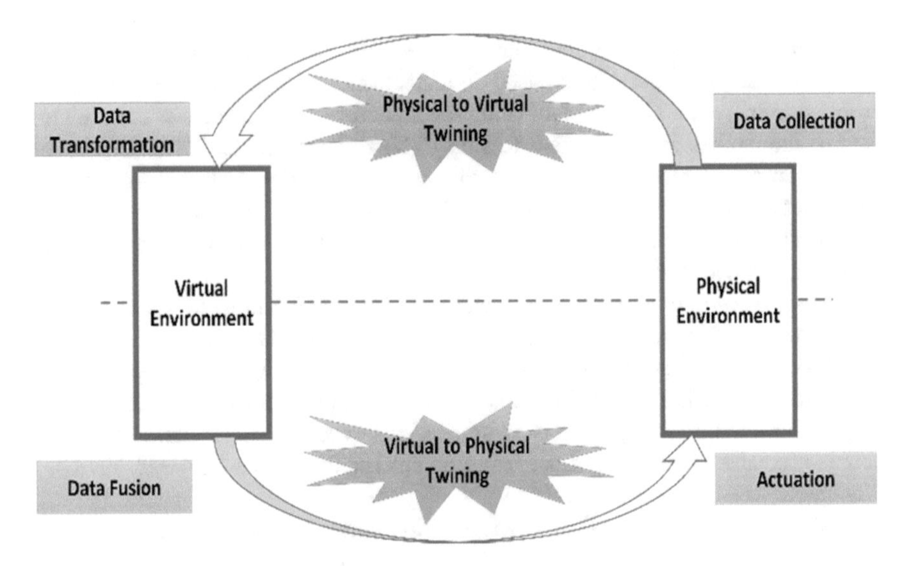

Fig. 2 The digital twining strategy for converting physical to virtual and vice versa

3 Methodologies

In this section, we attempt to develop a new VDR system that contains the following—assisting the customer in buying clothes through the internet by observing a model with a 3D characteristic and selecting the preferred color from the different package of colors, and using the email id and password to enable the users to authenticate and access the 3D VDR account—establishing an online Shopping cart to enable the customer to shop and check out the cart. Enabling secure registration and authentication and providing management facilities for customer's profiles and browsing through the electronic mall to determine the category of items required, such as shirts, dresses, pants, Caps, and Jackets. We are building a secured process for checking out from the shop, i.e., credit card verification. Furthermore, in this chapter, we proposed a quick size algorithm to assist the user in determining the prices and appropriate size. We utilized the basic four steps of the software development life cycle (SDLC) in this work, including analysis, design, implementation, and testing.

3.1 Analysis Stage

In the suggested VDR, we begin with the analysis stage, the system development life cycle (SDLC), during which present systems are evaluated, and alternative systems are offered. The analysis is divided into many phases. The first sub-phase entails identifying system requirements. As part of this sub-phase, any existing manual or computer systems that may be changed or upgraded as part of the project are carefully examined. Next, examine the needs and create them based on their interrelationships. Examine the requirements to determine their interrelationships, then create them. Create alternate first concepts to fulfill the requirements as a third step: decide which option best fits the organization's needs. Evaluating the alternatives and establishing which one has the lowest cost, labor, and technology level commitment is necessary. To examine the system, we first create a context diagram, followed by a data flow diagram, and last, a sample of the data dictionary.

3.2 Design Stage

The analysis turns the proposed alternative's description into the physical and logical specifications of the system during the system design process. We should design everything about the system, from input and output displays to reports, databases, and IT procedures. The logical design is not connected to any system hardware or software platform; instead, it concentrates on the system's business characteristics, such as how it will influence the organization's functional units.

Logical thinking is transformed into specifications or physics in the physical design phase. For example, you need to transform the original mapping, flow, and data processing diagrams in a system into a structural system design that can subsequently be broken down into smaller and smaller parts such that computer code may be generated from them.

To begin with, they must select which programming language the computer should be built in, which database system and file structure to use for the data platform, and finally, which operating system to use and which network environment to employ.

3.3 Implementation Stage

In the implementation stage, we utilized Web Graphics Library, commonly named WebGL, a JavaScript API for providing interactive 2D and 3D graphics, contained by any consistent web-browsers, not including plug-ins. In addition to WebGL, we utilized Three.js.

JavaScript framework for generating 3D material on the web allowing users to create models such as games, music videos, visualizations, and scientific data straight from their smartphone [25]. Blender is also a free and open-source 3D computer graphics software toolkit for creating animations, visual effects, art, 3D printed models, animation visuals, 3D compatible apps, and computer games [26], and 3D modeling and UV unpacking; texture development; raster graphics editing and skinning; particle simulation; mollusk modeling; sculpting; animation; motion joints; rendering; motion graphics; video editing and composition; and mollusk modeling.

However, Cascading Style Sheets (CSS) is essential for 2D and 3D modeling. CSS is a style sheet language used to specify the appearance of a document published in HTML or XML (including XML dialects such as SVG, MathML, or XHTML). As the name implies, CSS defines how components will be presented on a computer screen and in print, voice, and other media. Also necessary are Syntactically Awesome Style Sheets (SASS), which provide variables, nested rules, mixing, and functions with a completely CSS-compatible syntax. To keep huge stylesheets structured and to exchange designs inside and between projects, Sass may be used.

As a final step, we employed the use of Gulp, a web development tool that can assist with a variety of front-end tasks, including (i) starting a web server, (ii) automatically refreshing the browser after saving files, (iv) using a preprocessor like SASS, and (v) optimizing resources such as JavaScript and CSS. Figure 3 illustrates the processes involved in implementing the VDR website.

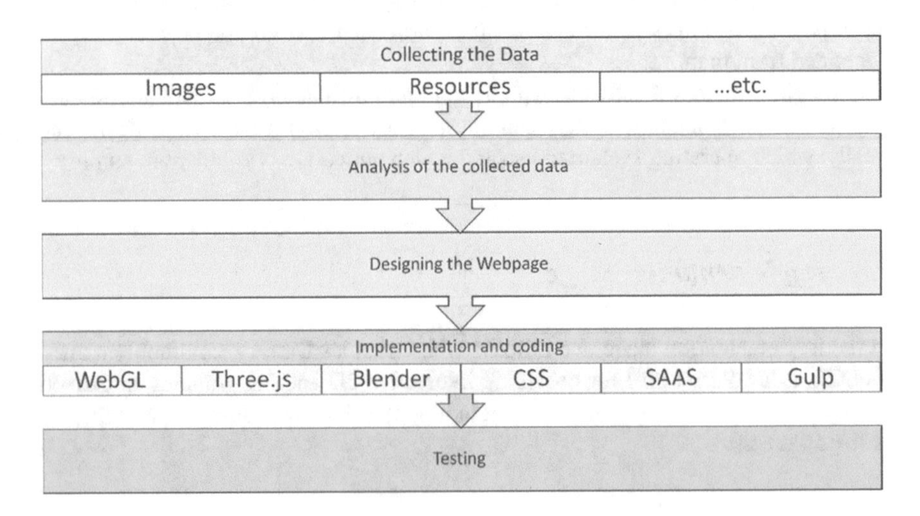

Fig. 3 The implementation of the VDR website

3.4 Testing Stage

System testing stage covers the examination of interaction points between units and subsystems. After unit testing of all units in the subsystems being tested is completed, system testing is performed. When units are combined, system testing identifies issues. Since the components are evaluated individually before they are combined (merged), any errors identified during the merging process are corrected.

Performance testing tracks and measures performance levels of the VDR under routine, low and high conditions. It assesses the resource consumption under the above parameters and serves as a basis for forecasting future resource requirements (if any). System testing aims to ensure that an application is accurate and complete in fulfilling its intended duties. System testing mimics real-world events in a "simulated real-world" test environment, and it tests all the system's functionalities that are needed in real life.

4 Experimental Results, Discussion, and Analysis

4.1 Collected Database Description and the Experiments Setup

The proposed VDR depends upon four basic tables that investigate the registration of Users, Category, clothing, and body area. The main fields of the users' table include four basic fields identifier (ID) as a primary key, username, password, and the image, as shown in Table 1. The category fields include the ID, name, description, area ID, and Image as investigated in Table 2. The VDR Clothing table fields include the ID, Name, Color Value, Price, Size ID, Category ID, Description, Image, and Gender, as shown in Table 3. The basic fields of the Body Area include ID, Area, and Description, as shown in Table 4. In this work, we utilized a front-end application to implement 3D-VDR based on CSS, SASS, Three.js, and Gulp. A created website consists of 25 app pages that contain some additional components to enable the users to select the suitable size of the clothes.

Table 1 The basic fields of users

Column name	Data type	Constraints
ID	Integer	Primary key
Username	Char (50)	
Password	Char (50)	
Image	Char (200)	

Table 2 The basic fields of category

Column name	Data type	Constraints
ID	Integer	Primary key
Name	Char (50)	
Description	Char (MAX)	
Area ID	Integer	Foreign key
Image	Char (200)	

Table 3 The basic fields of clothing

Column name	Data type	Constraints
ID	Integer	Primary key
Name	Char (50)	
Color Value	Char 50)	
Price	Integer	
Size ID	Small Integer	Foreign key
Category ID	Integer	Foreign key
Description	Char (MAX)	
Image	Char (200)	
Gender	Bit	

Table 4 The basic fields of the body area

Column name	Data type	Constraints
ID	Integer	Primary key
Area	Char (50)	
Description	Char (50)	

4.2 Authentication Using RSA

RSA is considered as one of the most common authentications to make the webpage more secure through the utilization of two-factor authentication-based Secure ID. RSA agent software is installed on the mobile server. The authentication agent intercepts requests for access to protected resources and directs them to the RSA authentication manager for authentication. As shown in Fig. 4, the Configuration to build a Secure ID for user authentication involves the construction of the RSA Manager, RSA Agent, and the Mobility Server. When the mobile client uses the secure ID to authenticate the user, the user enters the RSA Secure ID username and password into the mobile login dialog. The user has to read the token from the device and enter it manually. A software token is a software application on a USB device or smart card that provides a software token. If the authentication mode is to authenticate the user only, the password will be sent to the server in clear text, so it is vulnerable to active security attacks, especially when transmitted over a wireless connection.

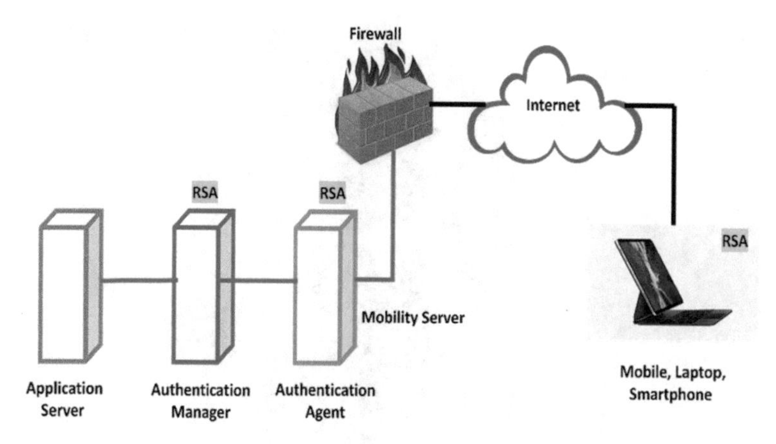

Fig. 4 The configuration of building secure authentication process based RSA

Using WPA or another protocol to encrypt data transmitted between the Mobility client and the wireless access point can prevent this vulnerability. WEP encryption of the wireless link cannot provide sufficient security, because the encryption key can be obtained by collecting sufficient encrypted data, but it is more secure than plain text transmission [27].

4.3 The VDR Webpage Components

The main components of VDR webpages consist of registering rooms. Secure authentication of the enrolled users to VDR is performed based on applying Email, First Name, Last Name, Password. They then register for the newly enrolled subjects. After the registration process, the store is loaded to check the different types of clothing in the store. If the customer chooses any clothing, the inner page category is checked in the selected clothing's different types, colors, and sizes. After that, the user can open his camera through the laptop or any smart device to take an image of himself that enables him to measure its different dimensions to choose the appropriate size, as shown in Fig. 5. In addition, the ability of the proposed VDR system to obtain the real-time user data, the Quick Size Algorithms assist the client in selecting the accurate size, and this algorithm performs for both males and females in a good manner, as shown in Fig. 6. To enable the client to change the selected color of the dress based on VDR, we present a customizer page that enables the client to change the colors of the 3D model of a selected product. It provides the visibility control of the product by rotating it 360 degrees and the zoom in/out to facilitate the selection process, as shown in Fig. 7. Another example of different poses and rotations controls based on the selected colors is further shown in Fig. 8 The Blender in the implementation stage describes the ability of the model to create something of your selection and apply the change of the 3D model after the authentication process, as

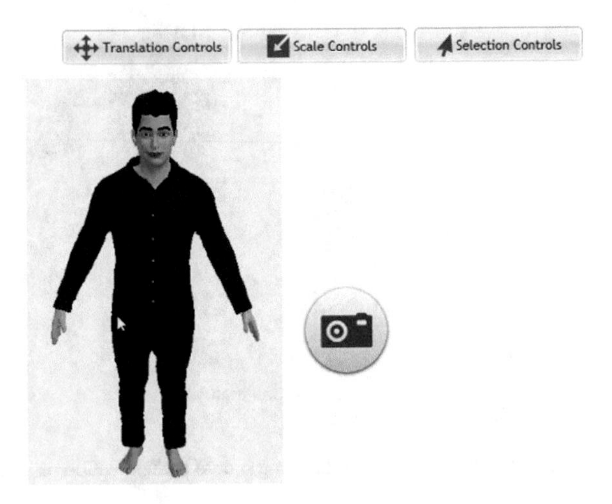

Fig. 5 The utilized camera is fixed inside the website

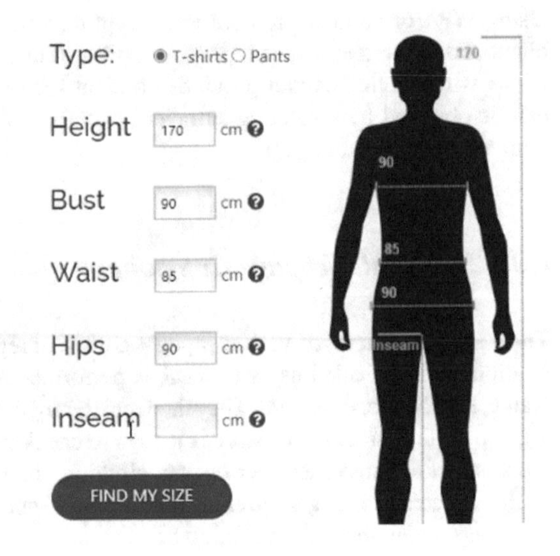

Fig. 6 The quick size algorithm for accurately measuring the clients dressing data sizes

shown in Fig. 9). Three.js Texture Loader is utilized to build a new texture, which generates a collection of texture repeats and sets packaging for it.

5 Conclusion and Future Work

A hybrid society of real and virtual reality is on the way. The rise of the metaverse has recently drawn significant attention from academia to industry. A metaverse is a network of three-dimensional (3D) virtual worlds designed to foster social interaction. People are physically isolated as a result of the coronavirus 2019 pandemic,

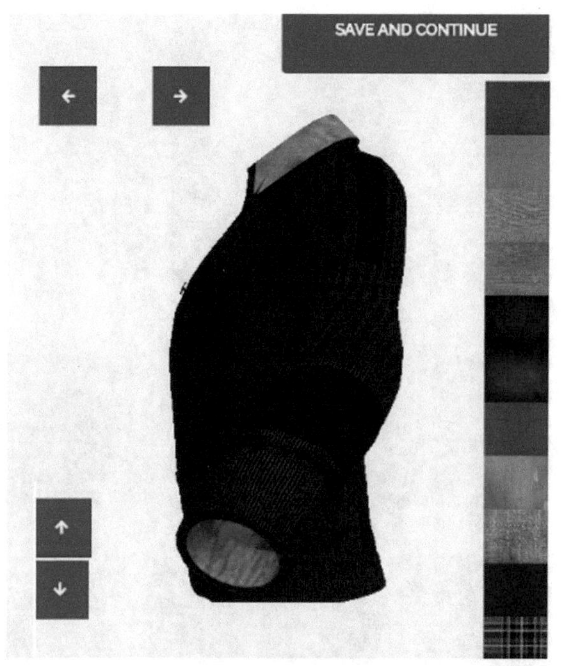

Fig. 7 An illustration of the chosen jacket's colors

which has fueled the growth of the metaverse. Unlike previous work, this commentary focuses on the metaverse's roadmap from the standpoint of artificial intelligence (AI). Digital twinning currently utilised to monitor the transaction process in E-Commerce and E-Shopping. This is because of its ability to transform physical reality into its equivalent digital reality. Through that, there are revolutionizing the industry by which complex models triggered by sensor updates and historical data can reflect almost every aspect of a product, process, or service. As in sustainable development, everything in the physical world will be replicated in the digital space through digital twin technology. In this paper, we implement a webpage using the most common tools to create a virtual dressing room that enables the clients to choose suitable clothes online easily and accurately. Furthermore, the suggested VDR system allows users to pick the optimum size and colors based on a 3D blender model. The proposed 3D VDR achieves the end price discriminant between the manufacturer and distributors.

Furthermore, the flexibility of the website and the ease of interaction are based on a traditional camera fixed on your laptop or any smartphone device. The response of the proposed 3D-VDR is very fast and precise. Sometimes for traditional users, online shopping is very complicated. Therefore, er present an easy graphical user interface that assists the user to select the appreciate clothes with a suitable color. In the future, we plan to expand the project to use in different stores such as foods and electronic devices based on meta-verse strategies.

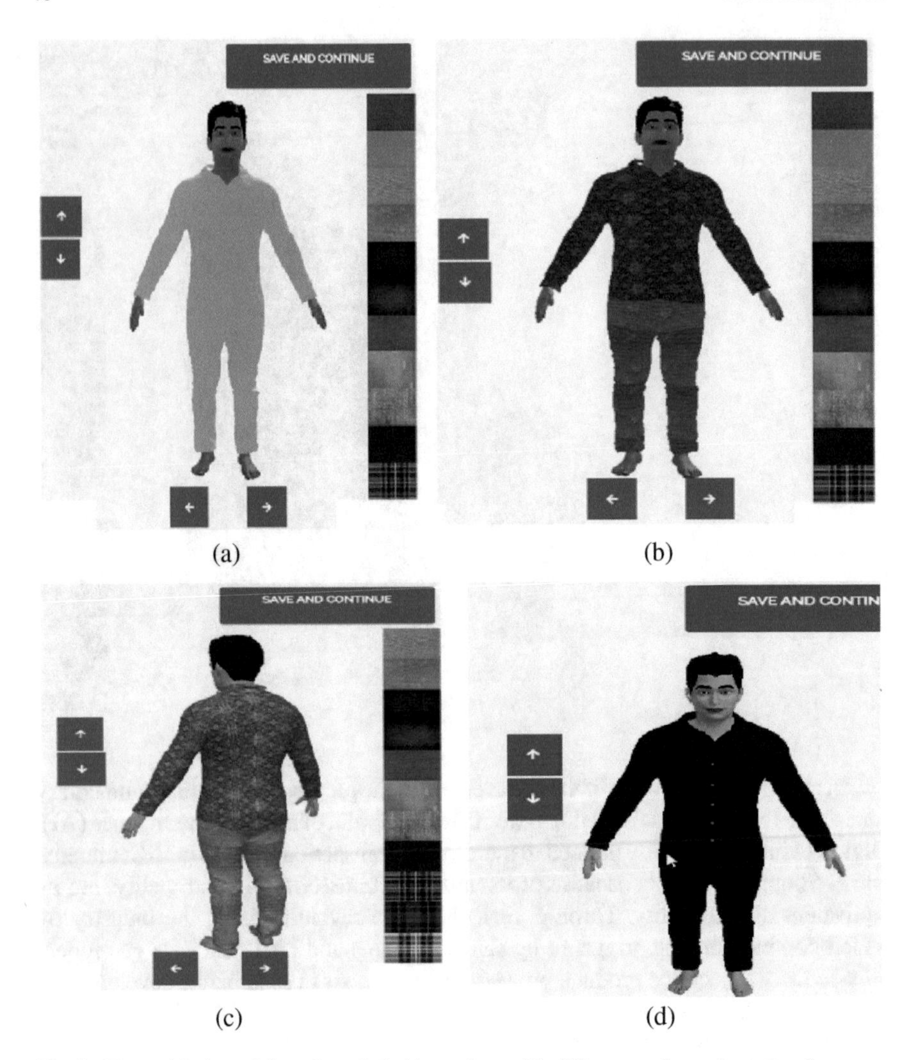

Fig. 8 The multi-view of the selected clothing colors with different angles and rotation degrees. **a** White color at frontal view, **b** Selection of preferred color frontal view, **c** Rotation view, **d** Black color and selection of the view and colors

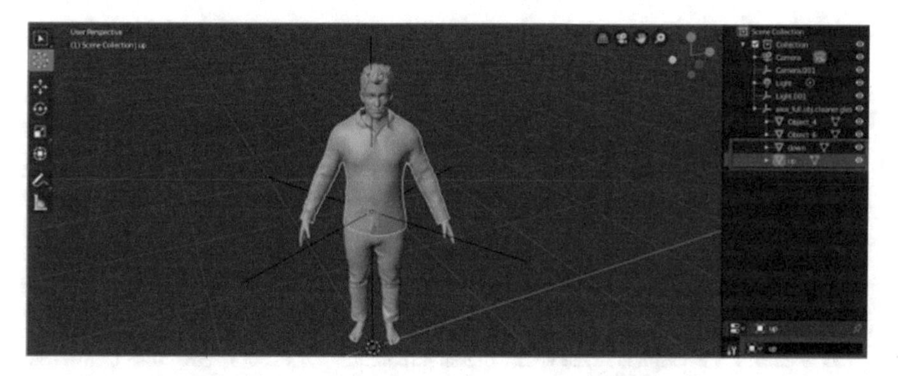

Fig. 9 The 3D model layer in blender

Acknowledgements We want to thank Misr Higher Institute for Commerce and Computers, especially Eng. Essam Nabil and Ahmed Bedawy, Amora Mahmoud, Basem Ghanem, Noura Al Hosainy, and Omar Fathy for their assistant in website implementation.

References

1. Liu, M., Fang, S., Dong, H., Xu, C.: Review of digital twin about concepts, technologies, and industrial applications. J. Manuf. Syst. **58**, 346–361 (2021)
2. Cliff Saran: The metaverse: A virtual reality check. ComputerWeekly.com. https://www.computerweekly.com/news/252515856/The-metaverse-A-virtual-reality-check. Accessed 10 Sep 2022
3. Bardzell, S., Shankar, K.: Video game technologies and virtual design: a study of virtual design teams in a metaverse. In: Virtual Reality, pp. 607–616. Berlin, Heidelberg (2007). https://doi.org/10.1007/978-3-540-73335-5_65
4. Uloziené, I., et al.: Subjective visual vertical assessment with mobile virtual reality system. Medicina (Mex.) **53**(6), 394–402 (2017). https://doi.org/10.1016/j.medici.2018.02.002
5. Szyjewski, G.: Conducting a secret ballot elections for virtual meetings. Procedia Comput. Sci. **192**, 4448–4457 (2021). https://doi.org/10.1016/j.procs.2021.09.222
6. Tsukamoto, T., Muroya, Y., Okamoto, M., Nakano, Y.: Collection and analysis of multimodal interaction in direction-giving dialogues: towards an automatic gesture selection mechanism for metaverse avatars. In: Agents for Educational Games and Simulations, pp. 94–105 Berlin, Heidelberg (2012). https://doi.org/10.1007/978-3-642-32326-3_6
7. Xue, F., Guo, H., Lu, W.: Digital twinning of construction objects: lessons learned from pose estimation methods. In: Proceedings of the 37th Information Technology for Construction Conference (CIB W78), São Paulo, Brazil, pp. 2–4 (2020)
8. Rathore, M.M., Shah, S.A., Shukla, D., Bentafat, E., Bakiras, S.: The role of ai, machine learning, and big data in digital twinning: a systematic literature review, challenges, and opportunities. IEEE Access **9**, 32030–32052 (2021)
9. Ahmed, M.: Anter, Yasmine S. Moemen, Ashraf darwish, aboul ella hassanien, multi-target QSAR modelling of chemo-genomic data analysis based on extreme learning machine, journal of knowledge-based systems, Elsevier. Knowl. Based Syst. **188**, 104977 (2020). https://doi.org/10.1016/j.knosys.2019.104977

10. Protopsaltou, D., Luible, C., Arevalo, M., Magnenat-Thalmann, N.: A body and garment creation method for an Internet based virtual fitting room. In: Advances in Modelling, Animation and Rendering, pp. 105–122. Springer (2002)
11. Al-Muhtadi, J., Ranganathan, A., Campbell, R., Mickunas, M.D.: A flexible, privacy-preserving authentication framework for ubiquitous computing environments. In: Proceedings 22nd International Conference on Distributed Computing Systems Workshops, pp. 771–776 (2002)
12. Isikdogan, F., Kara, G.: A Real Time Virtual Dressing Room Application using Kinect, p. 3 (2012)
13. Kantawong, S.: Development of RFID dressing robot using DC servo motor with fuzzy-PID control system. In: 2013 13th International Symposium on Communications and Information Technologies (ISCIT), pp. 14–19 (2013)
14. Priyadharsun, S., Lakshigan, S., Baheerathan, S.S., Rajasooriyar, S., Rajapaksha, U., Harshanath, S.B.: Parade in the virtual dressing room. In: 2018 13th International Conference on Computer Science & Education (ICCSE), pp. 1–4 (2018)
15. Kusumaningsih, A., Kurniawati, A., Angkoso, C.V., Yuniarno, E.M., Hariadi, M.: User experience measurement on virtual dressing room of Madura batik clothes. In:2017 International Conference on Sustainable Information Engineering and Technology (SIET), pp. 203–208 (2017)
16. Yaoyuneyong, G., Foster, J.K., Flynn, L.R.: Factors impacting the efficacy of augmented reality virtual dressing room technology as a tool for online visual merchandising. J. Glob. Fash. Mark. **5**(4), 283–296 (2014)
17. Noordin, S., Ashaari, N.S., Wook, T.S.M.T.: Virtual fitting room: the needs for usability and profound emotional elements. In: 2017 6th International Conference on Electrical Engineering and Informatics (ICEEI), pp. 1–6 (2017)
18. Kang, J.-Y. M.: Augmented reality and motion capture apparel e-shopping values and usage intention. Int. J. Cloth. Sci. Technol. (2014)
19. Liew, J.S.Y., Kaziunas, E., Liu, J., Zhuo, S.: Socially-interactive dressing room: an iterative evaluation on interface design. In: CHI'11 Extended Abstracts on Human Factors in Computing Systems, pp. 2023–2028 (2011)
20. Zhang, M., Wang, Y., Zhou, J., Pan, Z.: SimuMan: a simultaneous real-time method for representing motions and emotions of virtual human in metaverse. In:Internet of Things – ICIOT 2021, Cham, pp. 77–89 (2022).https://doi.org/10.1007/978-3-030-96068-1_6
21. Yaoyuneyong, G.S., Pollitte, W.A., Foster, J.K., Flynn, L.R.: Virtual dressing room media, buying intention and mediation. J. Res. Interact. Mark. **12**(1), 125–144 (2018). https://doi.org/10.1108/JRIM-06-2017-0042
22. Adikari, S.B., Ganegoda, N.C., Meegama, R.G., Wanniarachchi, I.L.: Applicability of a single depth sensor in real-time 3D clothes simulation: augmented reality virtual dressing room using kinect sensor. Adv. Hum.-Comput. Interact. (2020)
23. Zak, M.: Augmented Reality Try-On Adoption in the Online Clothing Industry: Understanding Key Challenges and Critical Success Factors, p. 67
24. Andronas, D., Kokotinis, G., Makris, S.: On modelling and handling of flexible materials: a review on digital twins and planning systems. Procedia CIRP **97**, 447–452 (2021)
25. Shi, W., Haga, A., Okada, Y.: Web-based 3D and 360o VR materials for iot security education and test supporting learning analytics. Internet Things, p. 100424 (2021)
26. Patoli, M.Z., Gkion, M., Al-Barakati, A., Zhang, W., Newbury, P., White, M.: An open source grid based render farm for blender 3d. In:2009 IEEE/PES Power Systems Conference and Exposition, pp. 1–6 (2009)
27. Dasgupta, D., Roy, A., Nag, A.: Multi-factor authentication. In: Advances in User Authentication, pp. 185–233. Springer (2017)

Metaverse for Brain Computer Interface: Towards New and Improved Applications

Sara Abdelghafar, Dalia Ezzat, Ashraf Darwish, and Aboul Ella Hassanien

Abstract The metaverse is a virtual-world human interaction. Virtual platforms can express human thoughts or even dreams, at least in the metaverse reality, but there are few restrictions on changing the surroundings or the avatars' appearance. A new era of human connection through the mind may be formed in the metaverse conditions when it is combined with the existing Brain-Computer Interface (BCI) technology, which permits system control via brain signals. BCI systems are intended to provide realistic and user-friendly communication channels based on brain impulses. This study examines BCI and metaverse technologies in depth, discussing the background of BCI and metaverse technologies as well as their individual various applications. The study then presents a review of works that have introduced the integration of BCI with Extended Reality (XR) -based applications, since XR has been viewed as an umbrella domain, covering Virtual Reality (VR), Augmented Reality (AR), and Mixed Reality (MR), which are the three key elements of metaverse technology. The study also goes on to discuss the future vision and possible applications that will combine BCI and metaverse technology together in the future.

Keywords Avatar · Augmented Reality · Brain Computer Interface · Extended Reality · Machine Learning · Metaverse · Mixed Reality · Motor Imagery · Virtual Reality · Virtual World

---

Sara Abdelghafar, Dalia Ezzat, Ashraf Darwish, Aboul Ella Hassanien: Scientific Research Group in Egypt (SRGE)

S. Abdelghafar (✉) · D. Ezzat
School of Computer Science, Canadian International College (CIC), Cairo, Egypt
e-mail: sara_abdelghafar@cic-cairo.com

D. Ezzat · A. E. Hassanien
Faculty of Computers and Artificial Intelligence, Cairo University, Cairo, Egypt

A. Darwish
Faculty of Science, Helwan University, Cairo, Egypt

A. E. Hassanien et al. (eds.), *The Future of Metaverse in the Virtual Era and Physical World*, Studies in Big Data 123, https://doi.org/10.1007/978-3-031-29132-6_3

1 Introduction

The Metaverse is an emerging technology that has the ability to revolutionize the world. The Metaverse is gaining popularity as a platform for human connection in the virtual world. The metaverse world can be reflected in human thoughts or even dreams via virtual platforms. The prefix "meta," which means "beyond," and the suffix "verse," which is short for "universe," are combined to form the term "Metaverse." As a result, it literally refers to an alternate universe. This "universe beyond" differs from metaphysical or spiritual ideas of realms beyond the material world since it is specifically a computer-generated universe. While "cyberspace" refers to the entirety of shared online space across all dimensions of representation, the term "metaverse" also refers to a completely immersive three-dimensional digital world [1–3]. Parallel to this, Brain-Computer Interface (BCI) is a technology that communicates human intent by decoding brain impulses associated with specific thoughts (e.g., visual imagery, imagined actions, or imagined speech). BCI refers to systems that enable brain-machine connection, which has a wide range of uses from healthcare equipment to human augmentation. Three main processes make up BCI's operation: gathering brain signals, deciphering them, and sending commands to a connected computer in accordance with the results. BCI can be used for many different things, such as neurofeedback, giving paralyzed people their motor abilities back, enabling communication with locked-in patients, and enhancing sensory processing [4–6]. Although the metaverse and BCI technologies may appear unconnected to one another, when carefully coupled, they could create a new generation of combined applications. This study gives insight into BCI and metaverse technologies, analyzing their background, types, components, main elements, as well as their numerous applications. The study then addresses the future vision and potential applications that will combine BCI and metaverse technology in the future.

The following are the study's primary contributions:

1. Provide an overview of BCI technology, its components, applications, and machine-learning roles.
2. Provide an overview of the metaverse, including its applications and use cases, and the three main components of the metaverse are Virtual Reality (VR), Augmented Reality (AR), and Mixed Reality (MR).
3. Review the current works which introduce BCI for Extended Reality (XR) applications.
4. Present the current and future fusion of Metaverse and BCI-based applications.

The rest of the study is organized as follows: Sect. 2 introduces an overview of BCI technology. Section 3 introduces an overview of Metaverse technology, applications and its main elements. Section 4 presents the literature review on works that combined BCI and XR-based applications. Section 5 discuss the current and future fusion of Metaverse and BCI-based applications. Finally, Sect. 6 concludes this work.

2 Brain-Computer Interface

A BCI measures brain activity, extracts features from that activity, and transforms those features into artificial output that replaces, enhances, supplements, or otherwise improves natural CNS output. As a result, the CNS's ongoing interactions with its internal or external environment are altered [7]. BCI could take the role of lost abilities like speaking or moving, such as regaining control over the body through stimulating the muscles or nerves that control the body. BCI has also been utilized to enhance functionality, including training users to enhance the grasp-required remaining functionality of damaged pathways. BCI can also improve performance, for example, by admonishing a drowsy driver to get up. The body's natural outputs may also be supplemented by a BCI, maybe through the use of a third hand [8].

The history of BCIs begins with Hans Berger's discovery of the electrical activity of the human brain and the creation of electroencephalography (EEG). Berger was the first to use an EEG to capture human brain activity in 1924 [9]. By examining EEG traces, Berger was able to recognize rhythmic activity like Berger's wave or the alpha wave. When first checked with monkeys in the late 1960s, it was discovered that impulses from a single cortical neuron can be utilized to drive a meter needle through the way of employing electrical nerve impulses as information carriers in person-computer communication [10]. Following that, systematic human experimentation started in the 1970s. In 1973, Vidal made the first serious attempt to assess the viability of employing electrical nerve impulses in a person-computer conversation that would allow computers to act as a prosthetic extension of the brain or for the purpose of controlling objects like prostheses [11]. Later, Elbert et al. [12] demonstrated in 1980 that individuals who had biofeedback sessions of slow cortical potentials in EEG activity were able to alter those possibilities to control the vertical motions of a rocket image moving across a television screen. Research on human BCI initially advanced slowly and was limited by computer technology and knowledge of brain physiology. On the other hand, the volume of peer-reviewed papers in this area during the last few years has shown that BCI research is currently developing at an extraordinarily rapid rate [13–16].

2.1 Brain Computer Interface Components and Applications

Invasive and non-invasive BCI Systems are the two main categories. Non-invasive techniques use sensors attached to the scalp to measure the magnetic field or electrical potentials produced by the brain. When tiny electrodes are inserted into the brain to track the neuron's activity, it is invasive. Semi-invasive is a term used to describe invasive procedures in which the electrodes are positioned on the exposed surface of the brain rather than directly into the cortex [17]. Brain activity captured using EEG, a non-invasive brain imaging technology, is the most frequently used input to BCI devices. Using passive electrodes attached on the scalp, EEG captures the

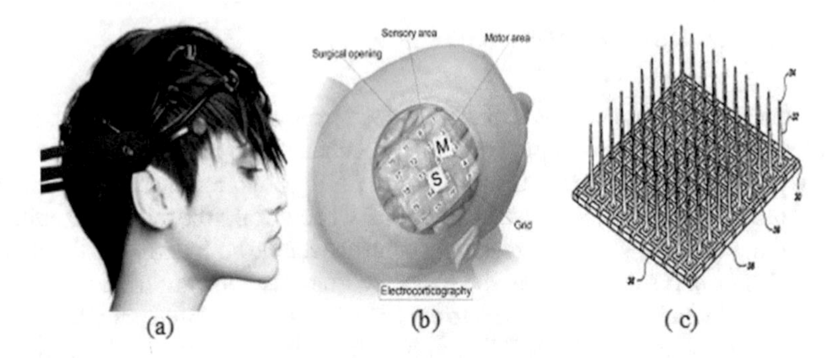

Fig. 1 Types of BCI: Non-invasive (a), Semi-invasive (b), and Invasive (c) [19]

electrical activity of the brain. Due to its portability, low cost, non-invasiveness, and greater time resolution ideal for real-time applications, EEG is more popular than other imaging techniques [18].

BCI establishes a direct communication channel between the brain and the outside world as well as a bi-directional communication interface between the brain and the outside world to link the human brain with peripheral equipment. BCIs offer a muscle-free means of communicating a person's intentions to outer/external gadgets like computers, brain prosthesis, and other aids. Unlike traditional input devices like the keyboard, mouse, and pen, the brain-computer interface reads signals produced by the human brain at specific locations and converts them into commands and actions that can be used to control one or more computers to carry out desired control/monitoring tasks BIC's input types are depicted in Fig. 1. There are three techniques for acquiring brain signals, as shown in the figure: non-invasive, which involves placing sensors on the scalp to measure electrical potentials produced by the brain or the magnetic field (MEG); semi-invasive, which involves placing electrodes on the exposed surface of the brain; and invasive, which involves inserting micro-electrodes directly into the cortex to measure the activity of a single neuron.

In general terms, the BCI system transforms brain impulses into comprehensible output commands to control external equipment. The brain sends out signals to control the user's intentions. BCI system typically consists of four parts: the acquisition of brain signals, signal processing—which includes preprocessing, feature extraction, and classification algorithms for classifying the brain signals—and transmission of the discovered commands to the controlling machinery and, finally, to the tool or device that will transmit back the feedback once the required action has been completed, as depicted in Fig. 2. The idea of a BCI is to identify and measure aspects of brain signals that represent the user's intents and to convert those aspects into device commands that carry out the user's intent in real time. Signal acquisition, which is the measuring of brain signals using a specific sensor modality, such as scalp or intracranial electrodes for electrophysiological activity, initiates this process. The signals are filtered to remove electrical noise and other undesired signal characteristics before being amplified to levels appropriate for electronic processing. The signals are then converted to digital form and sent to a computer. The second phase in interpreting brain signals comes after signal acquisition. Preprocessing, feature extraction, and classification are the three main operations of the processing unit. In preprocessing, to achieve a reliable assessment of the brain signal characteristics, noisy, environmental and physiological artifacts, such as electromyography signals, are avoided or eliminated.

Then, in order to identify relevant signal characteristics (i.e., the signals are ought to be highly correlated with the user's intention) from unimportant information and concisely express them so they can be transformed into output commands. The process of removing the important features from digital signals is known as feature extraction. The generated signal features are then converted by the feature classification method into the appropriate commands for the output device (i.e., commands that accomplish the user's objective). To accommodate and adapt to unanticipated

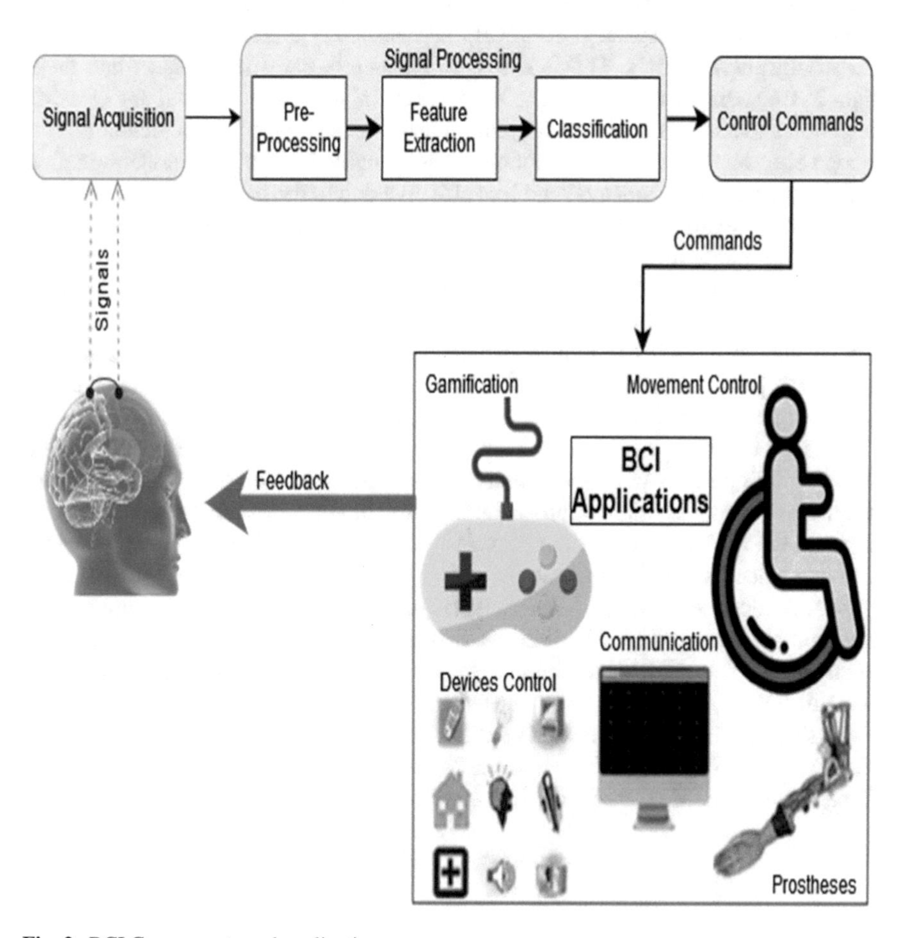

Fig. 2 BCI Components and applications

or learnt changes in the signal characteristics and to ensure that the user's prospective feature value range covers the whole range of device control, the classification algorithm should be dynamic. The commands from the classification algorithm drive the external device and perform operations like control in devices, movement, prostheses, communication, gamification, and so more. The control loop is closed when a user receives the feedback from device operation [20, 21].

2.2 Machine Learning Techniques for Brain Computer Interface

Practical BCI applications face a number of challenges. The EEG or other brain waves are essential in BCI systems for monitoring, regulating, and detecting human behavior. Many BCI systems, however, have difficulty effectively identifying and classifying these signals. The BCI system frequently performs worse when these signals are wrongly classified. The complexity of BCI systems used for emotion detection and mental condition recognition is increased by the lack of techniques for recognizing mental states and emotions. Additionally, human motion identification in BCI systems is complicated and less effective due to the lack of enough accurate methods for identifying the hand/limb motions of specific users. The techniques used for feature extraction, selection, classification, and detection of recognition characteristics are the foundation for all of these BCI tasks. Thus, by improving these algorithms, BCI systems and their detecting capabilities can be improved. Therefore, using machine learning techniques for feature extraction, selection, and classification can help improve the performance of BCI systems and produce better results, enabling BCI to cope with practical difficulties more successfully and effectively.

This has inspired researchers to look into more effective techniques for evaluating and completing BCI activities with more accuracy and precision as well as to look into how these techniques might improve the functionality of current BCI applications. As depicted in Fig. 3, BCIs employ various machine learning methods for feature extraction and selection, including Autoregressive (AR), Principal Component Analysis (PCA), Independent Component Analysis (ICA), Common Spatial Pattern (CSP), Fast Fourier Transform (FFT), Wavelet Transform (WT), and others. Then, in order to classify signals, BCI systems employ a variety of classifier algorithms, including Support Vector Machine (SVM), K-Nearest Neighbor (KNN), Linear Discriminant Analysis (LDA), Naive Bayes, Extreme Learning Machine (ELM) , Artificial Neural Networks (ANN), Convolutional Neural Networks (CNN), Logistic Regression, and Long Short-Term Memory (LSTM) [22–24].

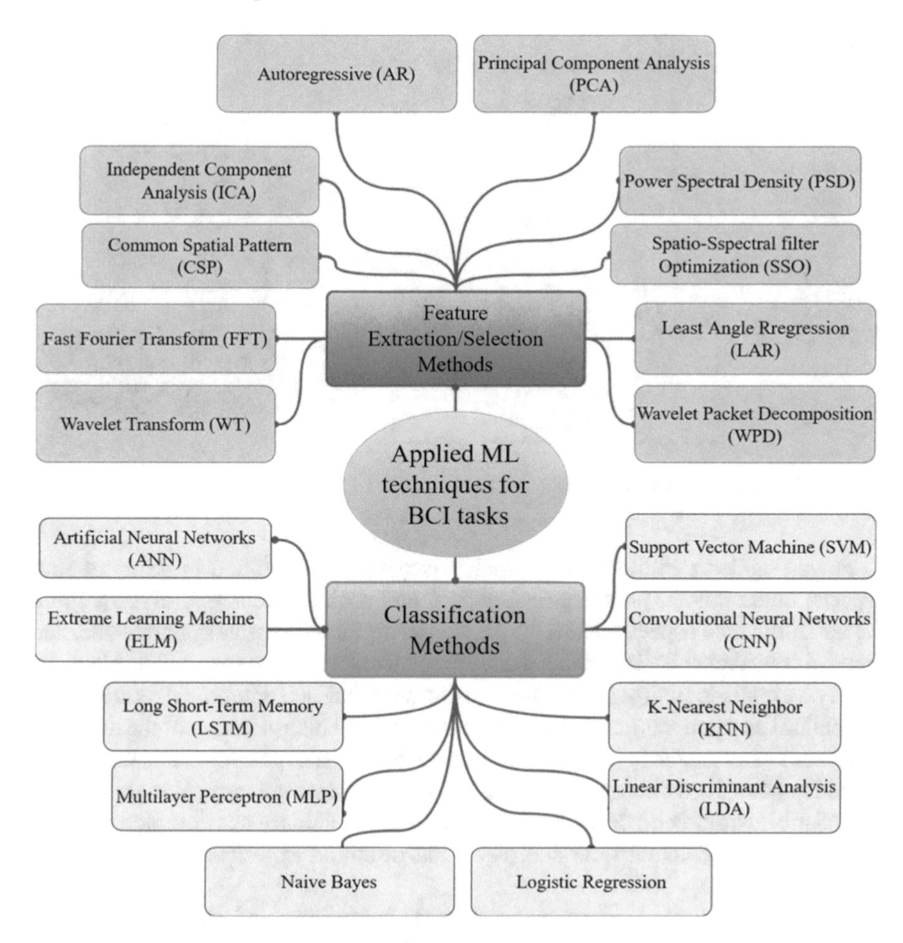

Fig. 3 Applied machine learning techniques for BCI process

3 Metaverse Technology

The idea of the Metaverse was floated three decades ago as a fictional story in which users are represented as avatars in disconnected virtual spaces, but until recently it came to prominence with the rebranding of Facebook to "Meta" [24]. The metaverse can be said to be a virtual environment that combines physical and digital and can be made possible by the fusion of the Internet and Web technologies, and Extended Reality (XR) [25]. XR, which includes virtual reality, mixed reality, and augmented reality as shown in Fig. 4, integrates the digital and physical worlds to varying degrees. In a similar way, the metaverse symbolizes the coexistence of the real world and a replica of digital environments in which each individual user has an avatar that resembles their physical selves and can experience a different life in a virtual world [26].

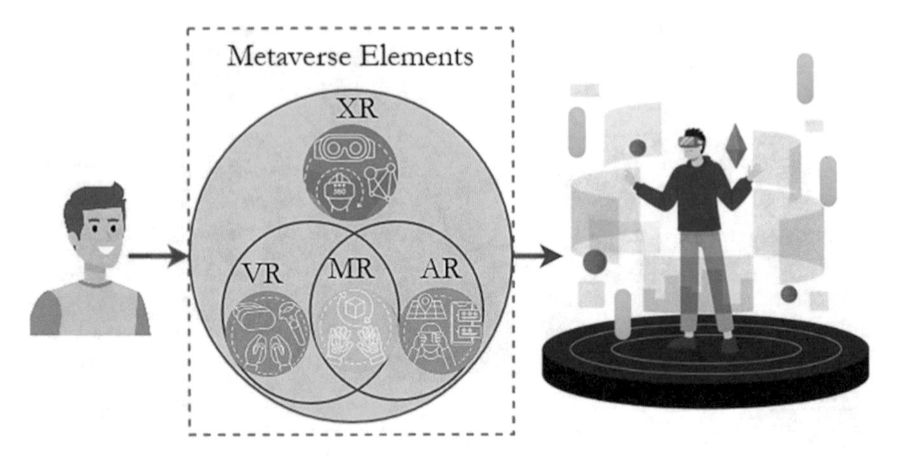

Fig. 4 The key elements of Metaverse

For a detailed definition of the metaverse, it can be defined in terms of four aspects: environment, interface, interaction, and social benefit [25, 27–29]. From the environmental aspect, Metaverse environments include realistic, unrealistic, and blended environments. The blended environment displays some unrealistic elements based on a realistic environment. The realistic metaverse accurately depicts the terrain and natural components according to the intention and interpretation of the designer.

Avatars cannot exist in two places at once in the realistic metaverse, and movement speed is constrained similarly to how it is in the physical universe. Whereas, in the unrealistic environment, the user is deceived by the imaginary world, which removes the constraints of actual time and place and offers an experience that cannot be replicated in reality.

From the interface perspective, there are 3D, immersive, and physical types. The 3D is characterized as giving a sense of realism, but it has a downside in terms of service continuity because it requires high performance hardware. The type of immersion is necessary to engage the user in the metaverse and maintain the existential state. To achieve immersion, a physical instrument (such as VR) is used to replace the user's natural visual sense. As for the physical type, although reversing the physical components in the interface is a good way to add realism, the level of realism that can be achieved with current technology is not enough. For instance, it can be challenging to convey tactile emotions with avatars such as hugs.

Social networking, personal dialogue, and communication are the three types of interaction that exist in the metaverse. The popularity of the metaverse can be increased by using social networks, which serve as the basis for connecting people's interactions. So the value of this technology will be created through collaboration rather than individual VR experiences. Personal dialogue preserves to appear natural because it reflects true personal qualities. The metaverse depends heavily on communication. User avatars are able to collaborate and exchange knowledge. Through this sharing and communication, they produce new value. This communication allows us to transcend time and space, unlike in the real world. It also gives users a reason to interact with one another and keeps the metaverse's society functioning. A new benefit can be added to society in the metaverse environment. Users can share their unique experiences and learn new things through the metaverse. By doing so, users increase their financial wealth, produce new things, and get the chance to display a different aspect of themselves. The metaverse can be used in various sectors such as gaming, education, tourism, distance working, healthcare, and banking, as shown in Fig. 5.

It can be said that the gaming sector is one of the biggest and most significant investors in metaverse technology. With the help of this technology, players can communicate with other participants in a single, open environment. The teaching and learning capabilities of Metaverse are as promising as its other applications. The learning environment will be raised to a new level of quality thanks to the utilization of virtual reality and the impacts of the Metaverse. With more intense and superior cognitive resources, students may observe live experiences. Virtual tourism is one of the metaverse's most common use cases. For those who are unable to travel over vast distances, the technology in this case allows for virtual travel. The biggest innovation in the travel sector can be attributed to the Metaverse, which enables the creation of immersive digital experiences with AR and VR. The metaverse will enable remote workers to improve their communication and place a greater emphasis on teamwork

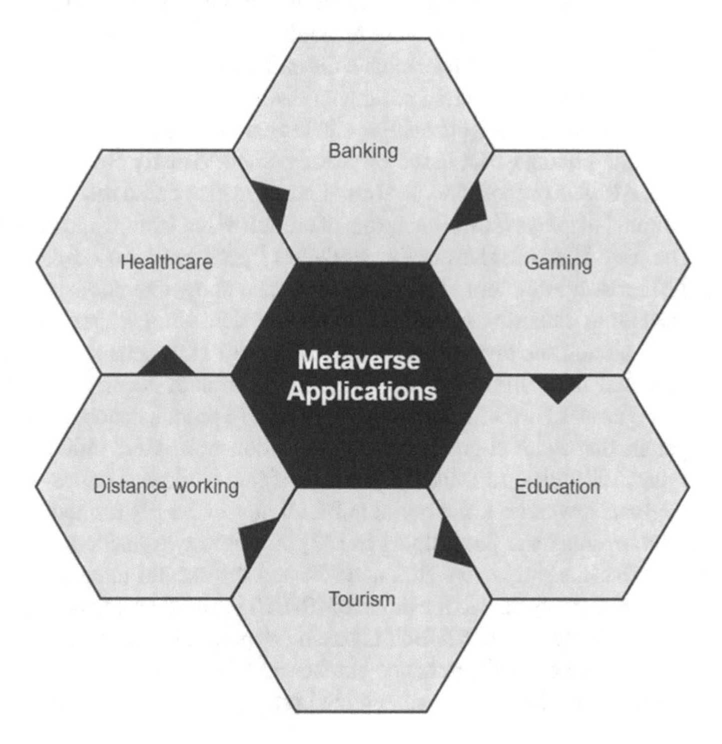

Fig. 5 Some Metaverse applications

through the use of virtual reality devices and 3D avatars. Metaverse can be used in healthcare to provide fully immersive medical experiences to patients and physicians.

The immersive virtual environment has the potential to generate new treatments for certain mental illnesses. According to some studies, a virtual reality environment can influence psychological issues such as depression, anxiety, cognition, and even social functions. Customers can interact with their banks in many ways in the metaverse. They can visit virtual bank branches to connect with bankers and receive advice [25, 30–33].

4 Literature Review of a Brain-Computer Interface for Extended Reality Applications

A literature search was performed using Google Scholar. The search term "Brain Computer Interface BCI and Extended Reality" was used on December 2, 2022. Extended reality was used in the search term because it is considered the umbrella term for virtual reality, augmented reality, and mixed reality. To narrow the search, the year of publication was limited to "2021" and "2022". Following this limitation, the first 100 research papers were retrieved for analysis. To guarantee relevance and diversity of research papers for brain computer interface and extended reality, the selected research papers were manually scanned by title, keywords, and abstract. When this could not be determined, the entire research paper was examined. Through this investigational search, seventeen distinct and related research papers were found and chosen for in-depth analysis. In [34], a fusion system called AR-BCI, which fuses single-channel BCI based on Steady-State Visually Evoked Potentials (SSVEPs) and AR was proposed to perform a performance comparison of different AR head-mounted displays (HMD) in terms of stimuli visualization and detection of SSVEPs. The used HMD devices in this study are Epson Moverio BT-350, Oculus Rift S, and Microsoft HoloLens. The results obtained show that choosing the right HMD for presenting flickering stimuli is important to achieving proper performance in general, and among the three HMDs used, Microsoft HoloLens and Oculus Rift S performed better than Epson Moverio BT-350 in terms of accuracy. In [35], the authors propose an AR HMD-based design to develop a practical, accessible, home-use BCI system that is an alternative communication option for individuals with neuromuscular disabilities. In [36], the purpose of this work was to design, implement, and evaluate a wearable and portable BCI that utilized ER for neurofeedback linked with motor imagery. The authors in [37] proposed two fusion systems were proposed, the first merging active BCI and XR and the second integrating passive BCI and XR. While the system in the passive XR-BCI is based on engagement monitoring, the system in the active XR-BCI is based on motion image detection. Due to the fact that they can be used to enhance human–machine interaction for users who are able to operate machines or in customized rehabilitation, both systems offer a wide variety of applications in the context of health 4.0. Another novel system was proposed in [38], which combines an existing AR HMD the Microsoft HoloLens 2, which uses a covert visuospatial attention (CVSA)-based EEG BCI technology to focus attention on specific regions of the visual field without making obvious eye movements.

5 The Fusion of Brain-Computer Interface and Metaverse Applications

BCI typically converts the user's intention or mental activity into output by translating the electrophysiological brain signals. BCI technology was initially created to give persons with severe motor limitations a way to communicate. Indeed, there have been significant advancements in these fields over the past few years, with a number of groups successfully controlling prosthetic limbs, wheelchairs, and spellers with BCI. Recently, its use has expanded beyond just medical applications to include both patients and healthy people, particularly in the fields of entertainment and multimedia. Additional communication channels might be made available by combining BCI and metaverse conditions. BCI and AR/VR technology combined have recently been seen as very promising. A system with immersive 3D graphics and real-time feedback that the user may engage with by using the BCI is typically how such a combination is created. This BCI-AR/VR combination has a promising future, which is apparent on two levels. BCI is viewed by the virtual reality community as a brand-new input tool that could fundamentally alter how people engage with the virtual environment (VE). Furthermore, BCI might be easier to utilize than conventional technology. Metaverse technology, on the other hand, also seems to be a beneficial tool for BCI research. In fact, virtual feedback can provide BCI users with richer and more inspiring input than conventional feedback, which is typically presented as a straightforward 2D design. Virtual feedback could therefore improve the system's learnability, shorten the learning curve for BCI skills, and improve the accuracy of mental state classification. The virtual environment can also be utilized as a flexible, affordable, and secure teaching and testing environment for BCI application prototypes. As a result, VE can be used as a starting point and then use BCI real-world applications. Figure 6 shows the framework for current and future fusion of BCI based metaverse applications.

There are many challenges to overcome while designing a BCI system, not the least of which is the issue of accurately differentiating brain signal patterns from each participant. There are various methods for enhancing BCI performance. The majority of studies use machine learning approaches to apply signal processing and classification elements, as we discussed in the previous section. However, by enhancing the user's control tactics and consequently by identifying fresh and effective mental activities to attain dependable control, BCI performance can also be enhanced. The most typical imaging task employed in BCI today is motor imagery (MI), which

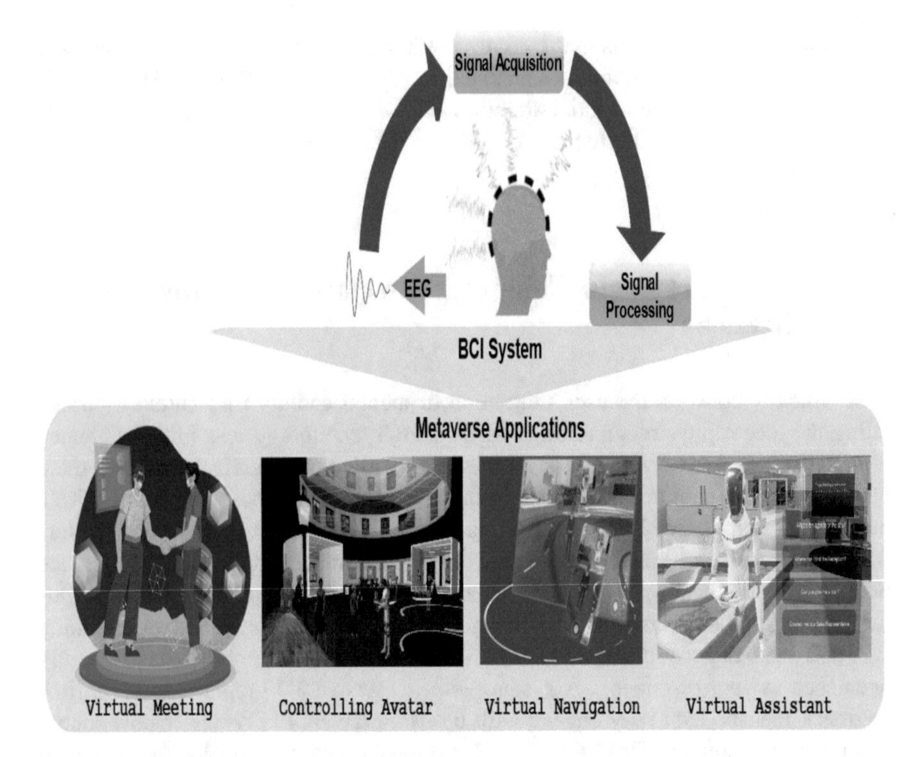

Fig. 6 The fusion of brain-computer interface and Metaverse applications

involves asking a user to imagine moving a body part. BCI with motor imagery (BCI-MI) is based on the classification and analysis of EEG patterns produced while imagining particular movements. The EEG over the sensorimotor cortex undergoes a distinctive alteration known as event-related de-/synchronization when different types of movements are imagined. The computer can use the user-specific patterns it has learnt to control movement. The same notion allows for simple movement control in a metaverse. Loeb and Pfurtscheller were the first to show the development and comparison of BCI-MI controlled VR by increasing the complexity of their investigations from controlling simple bar feedback in a synchronous manner to a self-paced (asynchronous) BCI in a highly immersive VE [39]. Then, numerous studies have demonstrated BCI-based VR applications employing MI in various virtual applications, including exploring a virtual pub [40], virtual street navigation [41, 42], investigating a virtual residence [42], investigating a digital library [43], virtually walking down a street [44], investigating the free space [45], visiting a park or maze [46, 47], and virtual museum visit [48]. Furthermore, several researchers have recently presented the use of BCI to operate an avatar in a metaverse environment utilizing imaging signals. These studies highlight the value of using virtual reality environments as feedback to stimulate BCI users and enhance their performances [49, 50].

To what extent can we distinguish between different mental processes of perceiving visual stimuli and picturing them in order to create BCIs based on EEG, there is another type of mental imagery, specifically imagined speech or visual imagery. Visual imaging is the modification of visual data that originates from memory rather than perception [51]. Imagined speech is the mental production of speech cues (phonemes, vowels, words, etc.) without auditory input or movement of the tongue, lips, or hand [52]. The future paradigm for intuitive BCI communication may use imagined speech or visual imagery. Future smart communication systems will include imagined speech and visual images, which may be broadly expanded to be modified for metaverse conditions. Potential metaverse uses can be seen in imagined speech and visual imagery, such as commanding virtual assistants in real-world communication systems like smart home systems or controlling avatars in metaverse environments. Additionally, the potential of visual imagery- and speech-based intelligent communication can be used in the virtual world and may act as a virtual training field for neurofeedback and practical BCI to improve the user's BCI performance [53–55].

6 Conclusions

This study presented the cutting-edge combination of BCI with metaverse and covered significant contributions in this intersection of research disciplines. It provides a detailed introduction to the history, principles, key elements, and applications of BCI and metaverse technology. The study examines earlier research on the integration of BCI and XR technology in various applications. Combining XR and BCIs is advantageous in settings that favor hands-free interaction, and future work in this domain may attempt to investigate this combination in a variety of use cases than the ones addressed in depth. This review also discusses applications based on BmuCI and the metaverse that would benefit from the combination of these two technologies. Furthermore, the study discusses some potential future advances that can be pursued in this union of BCI and metaverse technology.

References

1. Hyun Lee, S., Eun Lee, Y., Whan Lee, S.: Toward imagined speech based smart communication system: potential applications on Metaverse conditions. 2022 10th International Winter Conference on Brain-Computer Interface (BCI), pp. 1–4 (2022). https://doi.org/10.1109/BCI53720.2022.9734827
2. Dionisio, J.D.N., III, W.G.B., Gilbert, R.: 3d virtual worlds and the metaverse: current status and future possibilities. ACM Comput. Surv. **45**(3), 1–38 (2013)
3. Luu, T.P., Nakagome, S., He, Y., Contreras-Vidal, J.L.: Real-time EEG-based brain-computer interface to a virtual avatar enhances cortical involvement in human treadmill walking. Sci. Rep. **7**(1), 88–95 (2017)

4. Hill, N.J., Wolpaw, J.R.: Brain–computer interface. Reference Module in Biomedical Sciences, Elsevier, (2016).https://doi.org/10.1016/B978-0-12-801238-3.99322-X
5. Allison, B.: Toward ubiquitous BCIs, pp. 357–387. Brain-Computer Interfaces, Springer (2009)
6. Leuthardt, E.C., Schalk, G., Wolpaw J.R., Ojemann, J.G., Moran, D.W.: A brain-computer interface using electrocorticographic signals in humans. J. Neural Eng. **1**(2), 63–71 (2004). https://doi.org/10.1088/1741-2560/1/2/001
7. Aliakbaryhosseinabadi, S., Jiang, N., Vuckovic, A., Lontis, R., Dremstrup, K., Farina, D., Mrachacz-Kersting, N.: Detection of movement intention from movement-related cortical potentials with different paradigms. Replace, Repair, Restore, Relieve- Bridging Clinical and Engineering Solutions in Neurorehabilitation, pp. 237–244. Springer (2014)
8. Schalk, G., Allison, B.Z.: Chapter 26—Noninvasive Brain–Computer Interfaces, Neuromodulation (Second Edition), pp. 357–377. Academic Press (2018). https://doi.org/10.1016/B978-0-12-805353-9.00026-7
9. İnce, R., Adanır, S.S., Sevmez, F.: The inventor of electroencephalography (EEG), Hans Berger (1873–1941). Childs Nerv. Syst. **37**, 2723–2724 (2021). https://doi.org/10.1007/s00381-020-04564-z
10. Fetz, E.E.: Operant conditioning of cortical unit activity. Science **163**(3870) (1969). https://doi.org/10.1126/science.163.3870.955
11. Vidal, J.: Toward direct brain-computer communication. Annu. Rev. Biophys. Bioeng., pp. 157–80 (1973). https://doi.org/10.1146/annurev.bb.02.060173.001105
12. Elbert, T., Rockstroh, B., Lutzenberger, W., Birbaumer, N.: Biofeedback of slow cortical potentials. I. Electroencephalogr Clin Neurophysiol **48**(3), 293–301 (1980). https://doi.org/10.1016/0013-4694(80)90265-5
13. Vaughan, T.M., Wolpaw, J.R.: The third international meeting on brain-computer interface technology: making a difference. IEEE Trans Neural Syst. Rehabil. Eng. **14**(2) (2006)
14. Shih, J.J., Krusienski, D.J., Wolpaw, J.R.: Brain-computer interfaces in medicine. Mayo Clin. Proc. **87**(3), 68–79 (2012). https://doi.org/10.1016/j.mayocp.2011.12.008
15. Alonso, N.L., Gomez-Gil, J.: Brain computer interfaces, a review. Sensors **12**, 1211–1279 (2012). https://doi.org/10.3390/s120201211
16. Kawala-Sterniuk, A., Browarska, N., Al-Bakri, A., Pelc, M., Zygarlicki, J., Sidikova M., Martinek, R., Gorzelanczyk, E.J.: Summary of over fifty years with brain-computer interfaces-a review. Brain Sci. **11**(1) (2021). https://doi.org/10.3390/brainsci11010043
17. Wolpaw, J.R., Birbaumer, N., McFarland, D.J., Pfurtscheller, G., Vaughan, T.M.: Brain–computer interfaces for communication and control. Clin. Neurophysiol, 67–91 (2002)
18. McFarland, D.J., Anderson, C.W., Muller, K.-R., Schlogl, A., Krusienski, D.J.: Bci meeting 2005-workshop on BCI signal processing: feature extraction and translation. IEEE Trans. Neural Syst. Rehabil. Eng. **14**(2), 135–138 (2006)
19. Neurotechedu: Intro to brain computer interface. http://learn.neurotechedu.com/introtobci/. Accessed 23 Dec 2022
20. Rao, R., Scherer, R.: Brain-computer interfacing. Signal Process. Mag. IEEE **27** (2010)
21. Hettiarachchi, T.I., Babaei, T., Thanh, N., Lim, Ch.P., Saeid, N.: A fresh look at functional link neural network for motor imagery-based brain-computer interface. J. Neurosci. Methods (2018). https://doi.org/10.1016/j.jneumeth.2018.05.001
22. Altaheri, H., Muhammad, G., Alsulaiman, M. et al.: Deep learning techniques for classification of electroencephalogram (EEG) motor imagery (MI) signals: a review. Neural Comput. Appl. (2021).https://doi.org/10.1007/s00521-021-06352-5
23. Robert, M.K., Krauledat, M., Dornhege, Curio, G.G., Benjamin, B.: Machine learning and applications for brain-computer interfacing. Lecture Notes in Computer Science, vol. 4557. Springer, Berlin, Heidelberg (2007). https://doi.org/10.1007/978-3-540-73345-4_80
24. Rasheed, S.A.: Review of the role of machine learning techniques towards brain-computer interface applications. Mach. Learn. Knowl. Extr **3**, 835–862 (2021). https://doi.org/10.3390/make3040042
25. Park, S.M., Kim, Y.G.: A Metaverse: taxonomy, components, applications, and open challenges. IEEE Access **10**, 4209–4251 (2022)

26. Lee, L.H., Braud, T., Zhou, P., Wang, L., Xu, D., Lin, Z., Kumar, A., Bermejo, C., Hui, P.: All one needs to know about metaverse: a complete survey on technological singularity, virtual ecosystem, and research agenda, arXiv preprint:2110.05352 (2021)
27. Dwivedi, Y.K., Hughes, L., Baabdullah, A.M., Navarrete, S.R., Giannakis, M., Al-Debei, M.M., Dennehy, D., Metri, B., Buhalis, D., Cheung, C.M., Conboy, K.: Metaverse beyond the hype: Multidisciplinary perspectives on emerging challenges, opportunities, and agenda for research, practice and policy. Int. J. Inf. Manag. **66** (2022).https://doi.org/10.1016/j.ijinfomgt.2022.102542
28. Zwart, M., Lindsay, D.: Governance and the global metaverse. In: Emerging Practices in Cyberculture and Social Networking, pp. 173–182 (2009)
29. Zhang, S., Dinan, E., Urbanek, J., Szlam, A., Kiela, D., Weston, J.: Per-sonalizing dialogue agents: I have a dog, do you have pets too? arXiv preprint:1801.07243 (2018)
30. Lee, C.W.: Application of metaverse service to healthcare industry: a strategic perspective. Int. J. Environ. Res. Public Health **19**(20) (2022). https://doi.org/10.3390/ijerph192013038
31. Falchuk, B., Loeb, S., Neff, R.: The social metaverse: battle for privacy. IEEE Technol. Soc. Mag. **37**(2), 52–61 (2018)
32. Yu, J.E.: Exploration of educational possibilities by four metaverse types in physical education. Technologies **10**(5) (2022). https://doi.org/10.3390/technologies10050104
33. Mystakidis, S.: Metaverse. Encyclopedia **2**(1), 486–497 (2022)
34. Arpaia, P., Benedetto, E.D., Paolis, D.L., D'Errico, G., Donato, N., Duraccio, L.: Highly wearable SSVEP-based BCI: performance comparison of augmented reality solutions for the flickering stimuli rendering. Measur. Sensors **18** (2021.)
35. Sahal, M., Dryden, E., Halac, M., Feldman, S., Patterson, T.H., Ayaz, H.: Augmented reality integrated brain computer interface for smart home control. International Conference on Applied Human Factors and Ergonomics, pp. 89–97. Springer, Cham (2021)
36. Arpaia, P., Coyle, D., Donnarumma, F., Esposito, A., Natalizio, A., Parvis, M.: Non-immersive versus immersive extended reality for motor imagery neurofeedback within a brain-computer interfaces. In: International Conference on Extended Reality, pp. 407–419. Springer, Cham (2022)
37. Arpaia, P., Esposito, A., Mancino, F., Moccaldi, N., Natalizio, A.: Active and passive brain-computer interfaces integrated with extended reality for applications in health 4.0. In: International Conference on Augmented Reality, Virtual Reality and Computer Graphics, pp. 392–405. Springer, Cham (2022)
38. Kosmyna, N., Hu, C.Y., Wang, Y., Wu, Q., Scheirer, C., Maes, P.: A pilot study using covert visuospatial attention as an EEG-based brain computer interface to enhance AR interaction. International Symposium on Wearable Computers, pp. 43–47 (2021)
39. Lotte, F. et al.: Combining BCI with virtual reality: towards new applications and improved BCI. Allison, Biological and Medical Physics, Biomedical Engineering. Springer, Berlin, Heidelberg (2012). https://doi.org/10.1007/978-3-642-29746-5_10
40. L´ecuyer, A., Lotte, F., Reilly, R., Leeb, R., Hirose, M., Slater, M.: Brain–computer interfaces, virtual reality and videogames. IEEE Comput. **41**(10), 66–72 (2008)
41. Leeb, R., Keinrath, C., Friedman, D., Guger, C., Scherer, R., Neuper, C., Garau, M., Antley, A., Steed, A., Slater, A., Pfurtscheller, M.G.: Walking by thinking: the brainwaves are crucial, not the muscles. Presence (Camb.) **15**, 500–514 (2006)
42. Leeb, R., Lee, F., Keinrath, C., Scherer, R., Bischof, H., Pfurtscheller, G.: Brain-computer communication: motivation, aim and impact of exploring a virtual apartment. IEEE Trans. Neural Syst. Rehabil. Eng. **15**, 473–482 (2007)
43. Leeb, R., Settgast, V., Fellner, D., Pfurtscheller, G.: Self-paced exploring of the Austrian National Library through thoughts. Int. J. Bioelectromagn **9**, 237–244 (2007)
44. Leeb, R., Friedman, D., Muller-Putz, G.R., Scherer, R., Slater, M.M., Pfurtscheller, G.: Self-paced (asynchronous) BCI control of a wheelchair in virtual environments: a case study with a tetraplegics. Comput. Intell. Neurosci **79** (2007)
45. Scherer, R., Lee, F., Schlogl, A., Leeb, R., Bischof, H., Pfurtscheller, G.: Towards self-paced brain–computer communication: navigation through virtual worlds. IEEE Trans. Biomed. Eng. **55**(2), 675–682 (2008)

46. Ron-Angevin, R., Diaz-Estrella, A., Velasco-Alvarez, F.: A two-class brain computer interface to freely navigate through virtual worlds. Biomedizinische Biomed. Tech. (Berl) **54**(3), 126–133 (2009)
47. Velasco-Alvarez, F., Ron-Angevin, R.: Free virtual navigation using motor imagery through an asynchronous brain–computer interface. Presence (Camb.) **19**(1), 71–81 (2010)
48. Lotte, F., Langhenhove, A.V., Lamarche, F., Ernest, T., Renard, Y., Arnaldi, B., Lecuyer, A.: Exploring large virtual environments by thoughts using a brain–computer interface based on motor imagery and high-level commands. Presence (Camb.) **19**(1), 54–70 (2010)
49. Cohen, O., Doron, D., Koppel, M., Malach, R., Friedman, D.: High Performance BCI in Controlling an Avatar Using the Missing Hand Representation in Long Term Amputees. Springer Briefs in Electrical and Computer Engineering. Springer, Cham (2019). https://doi.org/10.1007/978-3-030-05668-1_9
50. Longo, B.B., Benevides, A.B., Castillo, J., Bastos-Filho, T.: Using brain-computer interface to control an avatar in a virtual reality environment. 5th ISSNIP-IEEE Biosignals and Biorobotics Conference: Biosignals and Robotics for Better and Safer Living (BRC), pp. 1–4 (2014). https://doi.org/10.1109/BRC.2014.6880960
51. Knauff, M., Kassubek, J., Mulack, T., Greenlee, M.W.: Cortical activation evoked by visual mental imagery as measured by functional MRI. NeuroReport **11**, 3957–3962 (2000)
52. Martin, S.P., Brunner, P., Holdgraf, C., Heinze, H.J., Crone, N.E., Rieger, J.: Decoding Spectrotemporal features of overt and covert speech from the human cortex. Front. Neuroeng, 1–15 (2014). https://doi.org/10.3389/fneng.2014.00014
53. Lee, S.H., Lee, M., Jeong, J.H., Lee, S.W.: Towards an EEG-based Intuitive BCI communication system using imagined speech and visual imagery. IEEE International Conference on Systems, Man and Cybernetics (SMC), pp. 4409–4414 (2019). https://doi.org/10.1109/SMC.2019.8914645
54. Hyun, S.L., Minji, L., Seong-Whan, W.L.: Functional connectivity of imagined speech and visual imagery based on spectral dynamics. 9th International Winter Conference on Brain-Computer Interface, IEEE (2021). https://doi.org/10.1109/BCI51272.2021.9385302
55. Kosmyna, N., Lindgren, J.T., Lécuyer, A.: Attending to visual stimuli versus performing visual imagery as a control strategy for EEG-based brain-computer interfaces. Sci. Rep. **8** (2018). https://doi.org/10.1038/s41598-018-31472-9
56. Bibri, S.E., Allam, Z., Krogstie, J.: The Metaverse as a virtual form of data-driven smart urbanism: platformization and its underlying processes, institutional dimensions, and disruptive impacts. Comput. Urban Sci. **2**(1), 1–22 (2022)
57. Pfurtscheller, G., Leeb, R., Keinrath, C., Friedman, D., Neuper, C., Guger, C., Slater, M.: Walking from thought. Brain Res. **1071**(1), 145–152 (2006)

The Future of Metaverse in the Virtual Era and Physical World: Analysis and Applications

Heba Askr, Ashraf Darwish, Aboul Ella Hassanien, and ChatGPT

Abstract The increasing diversity of reported user experiences and the expanding technical capabilities of virtual worlds has led to a splintered definition of what a virtual world is and is not in the academic literature. This chapter defines the term "Metaverse," explains how it came to be, sets it apart from similar concepts, and details its many potential uses. When the actual and the virtual worlds collide, as they do in Augmented and Virtual Reality (AR and VR), the resulting technology is known as Mixed Reality (MR), a subset of VR-related technologies. Also, in this chapter, some important applications based on Metaverse are presented. Ethical issues that are related to the Metaverse are highlighted.

Keywords Metaverse · Augmented Reality (AR) · Virtual Reality (VR) · Mixed Reality (MR) · Internet of Body (IoB) · Virtual world

1 Introduction

Even though the term "metaverse" has only been in use by tech commentators and academics for a short time, it was first used in a novel by Neal Stephenson called "Snow Crash" in 1992. In the book, the metaverse is shown as a VR space where avatars and software agents use the internet and AR [1]. The metaverse is defined as a new version of the internet that combines the real and virtual worlds using VR

Heba Askr, Ashraf Darwish, Aboul Ella Hassanien: Scientific Research Group in Egypt (SRGE)

H. Askr (✉)
Faculty of Computers and Artificial Intelligence, University of Sadat City, Sadat City, Egypt
e-mail: heba.askr@fcai.usc.edu.eg

A. Darwish
Faculty of Science, Helwan University, Cairo, Egypt

A. E. Hassanien
Faculty of Computers and Artificial Intelligence, Cairo University, Cairo, Egypt

ChatGPT
Open Artificial Intelligence, OpenAI L.L.C, 3180 18Th Street, San Francisco, CA 94110, USA

A. E. Hassanien et al. (eds.), *The Future of Metaverse in the Virtual Era and Physical World*, Studies in Big Data 123, https://doi.org/10.1007/978-3-031-29132-6_4

headsets, blockchain technology, and avatars. Immersive and interactive multimedia-style online games have been around for a few years. Users can engage with other people in a virtual world using VR goggles and avatars. With the usage of VR haptic gloves, headsets, AR, and Extended Reality (XR), users can completely experience high degrees of engagement and immersion. Organizations are beginning to analyze the metaverse's potential.

Various metaverse-based applications (e.g., Roblox and ZEPETO) have recently received a lot of attention. The present metaverse differs significantly from the previous Second Life metaverse in four ways. (1) Thanks to progress in deep learning, the new metaverse is more lifelike and engaging than its predecessor. It also has superior recognition performance and a more natural generation model. (2) The modern metaverse, as opposed to the one focused on personal computers, takes use of mobile devices to increase accessibility and continuity. (3) Security technologies like blockchain and virtual currency have increased the efficiency and reliability of the metaverse's financial services (e.g., Dime, Bitcoin). (4) The limitations of real-world social activities have contributed to the rise in popularity of online alternatives (e.g., Covid-19). Immersion levels (e.g., 3D, VR) are also used to classify the metaverse's interface. Users of VR hardware can experience a whole new degree of immersion in a 3D world thanks to the metaverse, but that's not all it has to offer. When defining a metaverse, it's important to consider not only the environment and the interface, but also the user and non-player interactions, which go beyond simple spoken communication (NPCs). Recently, the metaverse has concentrated on redefining of the metaverse's social significance, rather than simply being a duplicate of real-world society.

This chapter is organized as follows. Section 2 presents the basics and background. Section 3 presents some applications based on Metaverse. Section 4 highlights some ethical issues when using Metaverse. Section 5 concludes this chapter.

2 Preliminaries and Background

2.1 Metaverse Concept

The Metaverse is a unified, digital society in which people can do all of their daily activities—from work to recreation to relaxation to commerce to socializing—without having to switch between their digital and physical selves. The metaverse is in its infancy, hence there is no one, all-encompassing definition that can be used. However, common conceptions of the metaverse and its potential are beginning to take shape. It's vital to note that several virtual worlds are emerging to help people expand and deepen their social relationships in digital spaces. Adding a third dimension to the web allows for more realistic and natural interactions with content. Access to essential commodities, services, and experiences may even be possible

Table 1 Web 2.0 versus Web 3.0

Category	Web 2.0	Web 3.0
Structure	• Owned by one entity • The maximization of shareholder value drives all decisions	• Community-based, typically with a decentralized autonomous framework for leadership • Users vote on major changes
Data storage	• Centralized	• Decentralized
Platform format	• PC • Hardware for creating and using virtual and augmented realities • Mobile/app	• Hardware for creating VR and AR experiences on a personal computer • There will soon be a mobile/app version
Payment's infrastructure	• Traditional methods of payment (such as credit/debit cards)	• Bitcoin Wallets
Digital assets ownership	• Leased on the same marketplace platform as the acquisition	• Tokens that cannot be exchanged for other assets (NFT)
Content creators	• Developers	• Community/developers

from the convenience of one's own home, thanks to the democratizing effects of the metaverse [2].

Establishing a foundation for both current metaverse features (those associated with Web 2.0) and future Web 3.0 features is essential. While we've attempted to highlight our thoughts on the key metaverse distinctions between the two in the table below, it's becoming increasingly difficult to draw a firm line between the two as more conventional virtual worlds adopt elements of blockchain-based worlds. Table 1 summarizes the key distinctions between Web 2.0 and Web 3.0 [3].

2.2 *The Adoption of Metaverse in the Physical World*

In the near future, a new company or famous person will announce their entry into the virtual world every day. While catchy headlines do play a role, a number of interrelated trends can also be seen. This metaverse dream is made possible by a confluence of several cutting-edge technologies. The quality and accessibility of AR and VR headsets have increased, leading to a better overall experience. Due to blockchain technology, virtual currencies and NFTs are now a practical reality. Tokens are being used to monetize artists' activity as new methods of transacting and owning digital products emerge. Besides monetization and value exchange, token holders can also take part in the administration of the site (e.g. vote on decisions). Digital goods and services are no longer limited to a particular game platform or brand, and this democratic ownership economy, in conjunction with the possibility of interoperability, might unleash vast economic prospects. From a sociological perspective, the proliferation of virtual worlds experiences is facilitating the development of communities based on shared ideals and the freer expression of individual identities. Meanwhile, COVID-19 has accelerated the digitalization of our lives by making prolonged and varied online participation and communication the norm. The precipitous emergence of the metaverse can be attributed to the convergence of scientific, social, and economic forces [4].

2.3 *Virtual World*

2.3.1 Virtual Reality

A virtual world is just one form of the virtual environment in which a user can interact. Users can explore many various sorts of virtual settings, such as simulations, and while these environments may appear to be world-like, they are not populated as previously stated. The term "multi-user virtual environment" (MUVE) is used interchangeably with "virtual worlds". Several academics have classified VR into three categories: (tele-)presence, interactivity, and immersion [5].

Each virtual world is a social space gathering thousands of users and allows them to engage in various social activities; (1) is a space within VR with potentials for discovery and filled with objects to interact with; (2) allows exploration and interaction mediated by a visual representation of a user who is a part of the world (an avatar); and (3) is multi-user in nature and (4) be persistent, which means that they do not disappear.

Software that uses potentially VR-based technologies is sometimes referred to as a "virtual environment" or "virtual world." Virtual worlds evolved from the first implementations of text-based role-playing games. Well-known examples of successful virtual worlds with hundreds of thousands of players include "Ultima Online," developed by Origin System Inc. in 1997, "Everquest," launched by Verant Interactive and Sony Online Entertainment in 1999. Similar to other MMORPGs, Second Life has struggled to bridge the gap between its user base and the rest of the community. Although high equipment prices and inadequate quality have long hindered its mainstream adoption, few people are aware that basic VR technology has been available since the 1960s. For decades, researchers from various fields, including computer science, engineering, and the social sciences, have been exploring VR technology. As VR technology has advanced rapidly, so has the market for it. Virtual reality headsets like Facebook's Oculus Quest and HTC's VIVE are poised to revolutionize gaming and entertainment, thus consumer applications, especially video games, make up a large chunk of the industry. VR technology was originally used for non-entertainment reasons, such as flight simulation and military training.

2.3.2 Augmented Reality

As opposed to a virtual world enhanced by the modeling of an image or video from the actual world in virtual things, which is what AV refers to, AR is the integration of physical and digital information through various technological tools, hence it augments the shown physical world with simulated data. Because of their intermediate position between the real and the virtual, AR and AV are together referred to as MR, which is defined as "a unique subset of VR-related technologies that involve the merging of real and virtual worlds."

MR is often used interchangeably with "interaction between digital and physical items" because of its broader meaning that encompasses both the virtual and physical realms. Finally, the term XR is occasionally used as a catch-all phrase for any of these technologies or to refer to their combined use; It's a term used to define the hybrid spaces where physical and digital elements coexist, as well as the interactions between humans and machines that result from the use of computers and wearables [6].

2.4 Internet of the Bodies

Figure 1 shows how the "Internet of Bodies" (IoB) could drastically alter our perspectives on self, health, and community. A user's location, vital signs, and even their thoughts can be recorded and stored by IoB devices. Infusion pumps and hospital beds with built-in sensors are only two examples of freestanding medical technology; additional examples include wearable devices like health trackers and prosthetics, and implantable gadgets like cardiac monitors and digital medications [7].

2.4.1 Artificial Pancreas

The artificial pancreas technology can integrate insulin pump technology and continuous glucose monitoring with the help of AI techniques and algorithms. This system can automate the insulin dose based on inputs from continuous glucose monitoring.

2.4.2 Implantable Sensors in the Body

Implantable sensors can provide expansive and more precise tracking of the human body than wearable sensors. Implantable sensors have another advantage such as a skin-grafted interface which can enable patients or users to control remotely other devices. In addition, in the future, clothes will contain sensors to record the temperature of the body.

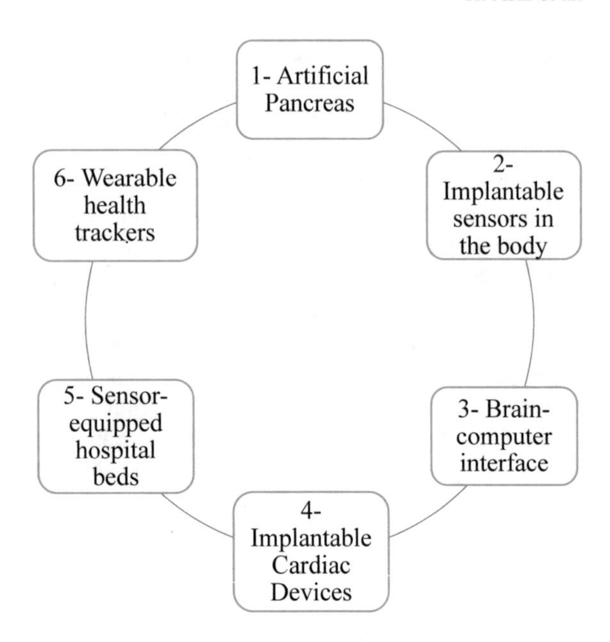

Fig. 1 Internet of body connectors

2.4.3 Brain-Computer Interface

The brain-computer interface can use electrodes to connect with signals from the brain to a computer system. They will be attached or wearable to the skull or implanted in the brain.

2.4.4 Implantable Cardiac Devices

Implantable cardioverter defibrillators, cardiac pacemakers and ventricular devices can provide continuous and real time information about the cardiac fluctuations of patients and remotely enable the device management to automate checks such as lead impedance, battery status, and sensing threshold.

2.4.5 Sensor-Equipped Hospital Beds

Beds in the hospitals and intensive care units will be equipped with sensors to measure some medical records of the patients such as blood pressure, temperature, heartbeat, oxygen and other medical records. Beds then can send these records to the central system of the hospitals using the internet of things to provide continuous real-time monitoring of the patients.

2.4.6 Wearable Health Trackers

Smartwatches rings smartphones and bracelets can be used to track heart rate, sleep patterns and other physical records. Such devices can provide displays and data analytics to provide information in accessible forms.

3 Metaverse Applications in Physical World

3.1 Medical Applications

Constraints placed on today's surgical trainees, such as time constraints and worries about patient safety, present a chance to improve upon tried-and-true teaching methods through the use of immersive technologies, such as virtual and augmented reality. Recent technological developments in user systems, a growing awareness of the need to update surgical education to meet the needs of today's trainees, and the realization that simulated environments may probably play a safe and cost-effective role in the application and mastery of surgical techniques and modalities are all contributing to the rise of AR and VR. According to this dogmatic approach, surgical residents had to put in a lot of time at the hospital before they could advance in their training and gain independence as surgeons. It is reasonable to question whether or not a modern surgical trainee can learn the skills necessary to perform the typical procedures that are fundamental to the practice of their specialization in a reasonable amount of time. The learner can learn the entire process in virtual reality, but without proprioceptive feedback, which may hinder their ability to apply what they've learned in the virtual world to real-world clinical situations.

3.1.1 Mixed Reality in Diagnostic Imaging

The diagnostic and preoperative imaging provided by radiology is crucial. In order to facilitate better operation preparation and monitoring, the authors of [8] suggest the first attempt to enhance radiological picture visualization by means of VR and AR. To aid in picture interpretation, modern workstations allow for the import of conventional radiological images in formats other than DICOM, their combination, and the development of three-dimensional reconstruction.

Therefore, it is crucial to enhance the precision of the segmentation in order to achieve the most precise amount of grey in each image pixel. One of the current difficulties in enhancing radiological images is the development of algorithms that can accurately perform the segmentation process automatically and generate a mesh of 3D models that can be imported into AR and VR systems for viewing and manipulation in the biomedical field. The goal of this effort is to improve the visualization of anatomical structures by bringing together VR/AR with MRI and other forms of digital imaging technology.

The new system incorporates three distinct technologies: (a) augmented reality, which displays 3D models of anatomical structures in real-world scale as if they had been printed in 3D; (b) virtual reality, which provides a complete immersion to study the 3D model with or without glasses, allowing even diving into the structure with hand movement recognition software; and (c) a desktop version, which allows users to use the system without mobile devices.

Surgeons who have experimented with the technique have found that it helps them better understand radiological results and can be a game-changer when it comes to preoperative planning.

All medical professionals in industrialized nations will visualize radiological results using these technologies within the next decade because of the low cost of AR glasses and other analogous devices and their pervasive use in society.

3.1.2 Virtual Reality for Human Anatomy

VR is becoming more common in our culture in a variety of settings, ranging from industry to entertainment. Since it allows the user to understand every structure of the human body by immersing themselves in each one, the authors of [9] are focusing on medical education, primarily anatomy, where its use is particularly intriguing. The goal of this research is to show how useful VR could be as a teaching aid in the field of medicine, specifically in the study of human anatomy. This method relies entirely on computer software that may be used with stereoscopic glasses to transport the user into a virtual world that displays an accurate representation of the skull and all of its bones and foramina, complete with descriptive audio.

3.1.3 Knee Arthroplasty Surgical Simulation and Education Using Virtual and Augmented Reality

XR technology let the surgeon to see the patient's anatomy in real-time, improving preoperative planning and offering intraoperative guidance. There are numerous potential applications for XR technology in orthopedic surgery. XR technology for orthopedic training and practice have showed considerable potential in allowing surgeons to visualize patient-specific anatomy in real time, improving preoperative planning, and providing intraoperative guidance to improve surgical intervention precision.

Modern computer and image technologies meet at the intersection of XR technologies such as immersive virtual reality (IVR), AR, and MR [10].

New XR technologies are becoming increasingly popular in many areas of modern medicine, but they are finding the most widespread application in surgical specialties due to their beneficial effects on patients' pre- and postoperative experiences. The medical field has been quick to adopt augmented and MR, and the operating room is no exception.

AR technology based on smart devices or head-mounted displays (HMDs) may be able to avoid the high capital costs associated with large-console navigation or robotic systems while obtaining comparable alignment precision in knee arthroplasty.

XR technology for orthopedic training and practice have shown considerable potential in allowing surgeons to visualize patient-specific anatomy in real time, improving preoperative planning, and providing intraoperative guidance to improve surgical intervention precision.

IVR technology has the potential to revolutionize surgical education and performance without increasing costs, leading to better patient outcomes.

While there is some evidence that IVR can help with short-term skill acquisition and transfer, more research is needed to determine whether or not it is also effective at ensuring long-term skill retention, improving the skills of experienced surgeons, coordinating the training of ancillary operating room staff, or being cost-effective.

3.1.4 Virtual Reality (VR) for Medical Education and Training

Current limitations continue to prevent the development of simulations that can distinguish the simulated reality from the real environment. According to certain research including 16 surgical residents, those who have been trained using VR techniques complete surgeries 29% faster than those who have been trained using traditional procedures, demonstrating yet another evidence of Virtual Reality's potential in this field [11].

3.2 Metaverse in Education

Developments in gaming hardware and software is a direct result of the industry's high demand from consumers, have had a significant impact on the creation of 3D simulated surroundings and bodies (e.g., virtual worlds and avatars) in recent years. In order to deliver dynamic and immersive learning experiences and boost learner engagement, 3D virtual world settings have the ability to serve as an educational tool for real-life simulations, professional training, synchronous interaction, and worldwide cooperation. In order to better understand how 3D virtual world environments can either encourage or undermine sustained engagement, thereby affecting users' experience and motivational behaviour, the authors of [12] utilised the study frameworks of Self-Determination Theory (SDT) and hedonic theory. This research adds to our knowledge of how students are influenced by their experiences in asynchronous, online learning environments.

By incorporating virtual world technology into the design of educational spaces, we can provide students access to novel online learning opportunities in a variety of subject areas and provide them with interactive features and simulation activities to supplement traditional classroom lectures. Professionals in the field of education say that the findings of this study are essential in describing the worth of 3D VR technology as an application that can improve students' motivation and learning experiences in the construction of a productive and engaging learning environment.

In order to better understand how different types of virtual experiences affect students' ability to learn, researchers in the future should employ an experimental methodology. Additional psychological components associated with actionable design factors of virtual environments could be the focus of future study if we gain more nuanced educational insights.

Over the past few decades, new theories of embodied and extended cognition have evolved to discuss the role of the body and environment in the cognitive process. Researchers in [13] looked into "virtually" embodied and extended cognition by studying two case studies of student interaction in a simulated language-learning environment. Two key findings from the study demonstrate the potential of 3D virtual worlds as sites for embodied and extended cognition: first, students frequently fail to differentiate between themselves and their avatars; and second, the boundary lines between the real and virtual environments are high (cognitively) permeable.

The creation of a 3D virtual environment is a time-consuming process that is tied to instructional strategies since it calls for a variety of advanced skillsets such virtual world design, creativity, problem-solving ability, and spatial thinking. The purpose of [14] is to investigate students' situational flow during 3D design creation in 3D MUVEs. This research aimed to examine the impact of prior situational flow experience within the context of a 15-week design education course.

The primary purpose of [15] was to investigate how an IVR science simulation affects learning in a university setting and whether or not self-explanation aids comprehension. After listening to the IVR biology lesson, the self-explanation group did a 10-min written self-explanation activity, whereas the control group simply unwound. Knowledge acquisition, procedure learning, and concept retention were all the same for the self-explanation group compared to the control. The results show that the IVR lesson was beneficial, but that supplementing it with a written self-explanation task did not boost retention after a lengthy IVR session.

3.3 Metaverse in Social Media Networks

Symbolic Interaction and Action The theoretical approaches that stress the Self as arising in context, through Self-Other and Self-environment interactions in the mundane details of daily life, likewise place an emphasis on face-to-face interactions rather than virtual ones.

Self-Other interactions take on a different hue in virtual worlds due to the nature of the communications and relationships that take place there; these interactions are hampered by a lack of social cues but are ultimately successful because of creative solutions developed in response to the desire to engage in social activity.

The authors of [16] argue that, while the birth of the self is interactive, virtual environments are separate venues for a Self in which the distinctive role of social interaction must be emphasized, and thus they place special emphasis on the role of symbolic mediation in this process. Avatars' physical, demographic, and personality qualities are discussed in [17], along with (2) how these representations shift across different online activity contexts, and (3) how there is variation in the assigned personality traits amongst participants.

3.4 Metaverse in Culture and Tourism

MR devices are used by many sectors, including the public sector and the entertainment industry, to develop interactive and compelling mobile applications. The museum setting is a perfect testing ground for MR navigation systems because of the flexibility afforded by spatial holographic Head-mounted Displays (HMD).

By merging interactive graphics from the past with actual exhibits and displays, MR systems at museums can enhance the standard museum visit. The majority of today's methods in MR guidance study focus on how visitors interact with particular materials. Virtual information systems are becoming important instruments for improving the museum visiting experience. The classic human tour guide system is a structured framework for engaging, amusing, and educating museum visitors with information along a defined itinerary. The goal of the research presented in [18] is to create a new type of tour guidance system that will enhance the museum-experience goer's without increasing the need for human tour guides. The data aggregation method looks into how well the MuseumEye app works with pre-existing pharaonic displays in a museum. The results of this research suggest that a majority of museum visitors see MR tour guides favourably, and that AR HMDs may be used effectively. This research suggests that MR technology can significantly enhance the museum-going experience for real-world visitors.

Museum tour suggestions can now be mined from the wealth of user-generated content present on social media regarding the quality, popularity, and reception of individual works of art and museum exhibitions. Data mining on information gathered by sensors placed in museums as part of the Internet of Things could be used to improve the quality of recommendations made to visitors. Sensors, augmented reality, and semantic technologies have been incorporated to revitalise the traditional museum experience.

Benefits for inference generation to enhance the museum experience for visitors can be derived from the semantic processing of this data (suggesting what artwork should tourists see first, or suggestions based on tourist profiles and analysed posts on social networks). Mobile technology, wireless connectivity, wearable sensors, and AR have transformed the museum visit from a routine of viewing artworks and making notes to an interactive and immersive experience. Various technologies and a vast amount of data generated daily on social networks, such as online photographs and videos and cultural tourism in museums that are experimenting and inventing, are continuously improving the user experience. In [19], the authors developed a cross-disciplinary strategy for recommending museum visits. This technique combines and integrates data sources to develop and recommend augmented reality-enhanced inside and outdoor itineraries for museums. The final product is an AR tour tailored to the user, complete with insider advice on how to get the most out of a trip to many locations.

3.5 Metaverse in Gaming and Entertainment Applications

Since its inception, the metaverse has had an impact on the advancement of science and technology. While Artificial Intelligence (AI) remains an important technology for solving challenges in the metaverse. Football instruction is an important aspect of physical education, and VR offers the qualities of immersion, interactivity, and imagination that can be used to create a virtual and realistic football teaching process.

From consoles to PCs and cellphones, gaming has long been an integral part of the Metaverse. Video games are becoming more realistic and visually appealing. Users will be able to enjoy amazing visuals and in-game gameplay. According to the three-stage Metaverse, the game-based Metaverse can provide countless opportunities. It could eventually evolve to virtual settings that are similar to real-world conditions. Roblox offers 50 million games and 3 billion hours of monthly usage. People are spending more time using social networking sites. Games are the most popular platform in the Metaverse. Aside from simply focusing on interests, there are ways to simplify difficult tasks by employing games. A blockchain-based game has gained in popularity as money and personal information are widely used in the Metaverse.

4 Ethical Issues in Metaverse

Several authors have analysed the moral ramifications of the virtual, arguing for a wide spectrum of viewpoints. These range from the view that the virtual is fundamentally nothing new and that old ethical theories apply directly to the polar opposite position that experiences in virtual worlds significantly alter our ability to act in the real world [20].

According to Floridi [21], we now spend the majority of our time inside the computer-based interactive environments that define our lives, relationships, and routines. The influence of our online behavior on our offline actions is one of the most pressing ethical concerns raised by virtual environments. Apps for VR can be developed to allow and visually simulate nearly any imaginable evil behavior, including murder, mutilation, torture, rape, robbery, and grand theft, contrary to popular belief.

4.1 *The Ethics of Smart Stadia: A Stakeholder Analysis of the Croke Park Project*

Research into the evolution of smart stadiums used the smart Croke Park stadium project in Dublin as a case study.

Data in vast quantities will be made available by smart technology on stadiums and their parts, visitors, energy use, and commercial potential. When someone collapses or gets hurt, when too many people are congregating in one area, when others are trying to sneak into the stadium, when a brawl or other disruption is likely to break out, and so on, smart technology will make it easier to warn the appropriate staff personnel.

The vast data sets collected by IoT technologies will be used by police and security services to forecast crowd behavior or the likely behavior of people whose information has been collected by the IoT. Such data will be useful outside of stadiums and should aid future improvements in predictive policing [22, 23].

The integration of sensor technologies in Smart Stadia compromises individual privacy by allowing users to control what is known about them [24, 25].

From the deontological perspective of ethical value promotion/violation and the consequentialist perspective of stakeholder value creation/destruction, several ethical opportunities and challenges associated with the development of the "Croke Park" smart stadium project in Dublin have been identified and discussed.

These technologies allow society to substantively investigate the values underlying present actions, including technical practices, as well as the prospect of building a new community of virtuous practice, whether focused at eudaimonia or some form of environmental flourishing. Analyzing the ethical problems and potential of technological innovation from the perspective of stakeholders gives a valuable foundation for resolving these complicated concerns in the first instance.

More research is needed into the rise of the surveillance society and the risks it poses to individual liberties and liberal democracies, as well as the accountability issues raised by the greener practices made possible by smart technology, the nature of sports, and the ownership of data.

4.2 *Analysis of the Moral Implications of MMORPGs' Game Mechanics and Technology Mediation*

When discussing the ethical issues of video games, the conversation frequently centers around the effects of its content, such as violence. This study contends that the effects of game mechanics, such as reward mechanisms, should also be studied, as these are at the heart of game attraction. The design of incentive mechanisms may influence how consumers interact with the game and how long they play. With this type of incentive addiction, the user is more likely to focus on something else when games are not available because the major factor is intrinsic to the player.

To learn how much predisposition for addiction is already there in players and what part the video game and its mechanics play, further research is needed on the association between motivation and attraction addiction, as well as a deeper look at users suffering from game addiction.

This research primarily addresses the issue of attraction addiction and its ethical implications within the framework of the game and the ethics of game design.

The game's fundamental mechanics could be designed with the concept of technology mediation in mind by examining the mechanics' mediating influence in the player's experience and actions while playing.

Part of the reason for bad playing patterns can undoubtedly be attributed to game mechanics' mediating effects: MMORPGs often use a random-ratio reinforcement schedule based on operant conditioning for goals and rewards.

Early accomplishments are virtually instantaneous, and advancement takes more and more time and effort until it is almost unnoticeable. The operant conditioning model lies at the center of behavioral game design in MMORPGs [26].

Most developers would be pleased if their MMORPG excels at retaining a user's attention for an extended period of time and entices gamers to continue playing. Although some Asian countries have imposed time limits on gaming, developers have yet to fully address the concerns raised by game mechanics, partly due to a failure to consider the consequences of their design choices from all perspectives.

Because this study did not aim to investigate the player's motivations or the context in which they engage in their problematic usage behaviors, we cannot say for sure what the long-term effects of such practices might be. Developers have the power and responsibility to improve the lives of millions of players by considering the ethical implications of their game's mechanics and implementing changes to encourage positive player behavior [27].

5 Conclusions and Recommendations

The idea of the Metaverse, a multi-user online space for collaborative and individual pursuits, has piqued people's imaginations for a long time. Recent technological developments, such as virtual and augmented reality, however, have greatly raised the possibility of the metaverse becoming a reality in both the virtual and physical worlds.

The metaverse has the potential to serve as a one-stop destination for all things done in the virtual sphere, including but not limited to, buying, socializing, and playing games. By allowing consumers to completely immerse themselves in these activities, virtual reality technology will provide a more exciting and authentic experience. As a result, we may see the rise of novel types of entertainment like online amusement parks and augmented reality concerts. Additionally, the metaverse is anticipated to develop into a stage for telecommuting, online learning, and the sale of virtual products and services.

Real estate, transportation, and retail are just a few of the actual industries that stand to benefit from the metaverse. Potential purchasers could, for instance, take advantage of virtual reality technology to take virtual tours of properties without having to physically visit them. The environmental toll of transportation could be lessened if the metaverse were utilised to simulate highways and other forms of physical infrastructure. The metaverse has the potential to revolutionise the retail industry by facilitating the development of immersive, online stores where customers may shop from any location.

In general, the metaverse will have far-reaching effects on the virtual and real worlds, opening up numerous new avenues for art, learning, and trade. However, the impact of the metaverse on real-world social relationships and other ethical and societal considerations should not be overlooked.

From a data scientist perspective, researchers can work on evaluating the data produced by metaverse activities to acquire insights into user behavior and preferences, as well as develop methods for understanding and modeling user behavior in the metaverse. We can also experiment with new ways to connect the virtual with the actual world, such as using augmented reality, and find out how to make the metaverse more user-friendly overall.

With the IoB, we can send and receive data in a shared virtual physical environment by being "tethered" to our digital counterparts, a concept that is central to the vision of a linked world driven by our bodies and identities.

As designers of interactive works, we were demonstrating to ourselves, through many public tests, that the living body was by far the most significant part of the mix, despite the common belief that it is the responsibility of technology to enable smooth interaction and connectedness. In this cooperative mixed realm of body and technology convergence, we start to develop the ability to move fluidly between virtual and physical existence. At now, approximately 200 different kinds of medical implants are available to patients. Although this is evolving quickly, currently most medical implants are not networked with the outside world. Metaverse is already being used in several practical contexts, and this trend is only expected to grow in the near future.

Acknowledgements A machine-generated summary based on the work of Girvan, Carina 2018 in Educational technology research and development

Thank you to all the software engineers and staff at OpenAI who developed the language model and AI chatbot known as ChatGPT and made it available open access (https://openai.com).

References

1. Joshua, J.: Information bodies: computational anxiety in Neal Stephenson's. snow crash. Interdisciplinary literary. Studies **19**(1), 17–47 (2017)
2. Cheong, B.C.: Avatars in the metaverse: potential legal issues and remedies. Int. Cybersecur. Law Rev. **3**, 467–494 (2022). https://doi.org/10.1365/s43439-022-00056-9

3. Pannicke, D., Zarnekow, R.: Virtual Worlds. Bus. Inf. Syst. Eng. **1**, 185–188 (2009). https://doi.org/10.1007/s12599-008-0016-1
4. Bray, D.A., Konsynski, B.R.: Virtual worlds, virtual economies, virtual institutions. Working Paper, Department of Decision and Information Analysis, Emory University, Atlanta. (2007) http://ssrn.com/abstract=962501
5. Wohlgenannt, I., Simons, A., Stieglitz, S.: Virtual Reality. Bus Inf Syst Eng **62**, 455–461 (2020). https://doi.org/10.1007/s12599-020-00658-9
6. Flavián, C., Ibáñez-Sánchez, S., Orús, C.: The impact of virtual, augmented, and mixed reality technologies on the customer experience. J. Bus. Res. 100:547–560 (2019)
7. Boddington, G.: The Internet of Bodies—alive, connected, and collective: the virtual physical future of our bodies and our senses. AI and Soc. (2021). https://doi.org/10.1007/s00146-020-01137-1
8. González Izard, S., Juanes Méndez, J.A., Ruisoto Palomera, P., et al.: Applications of Virtual and Augmented Reality in Biomedical Imaging. J. Med. Syst. **43**, 102 (2019). https://doi.org/10.1007/s10916-019-1239-z
9. Izard, S.G., Juanes Méndez, J.A., Palomera, P.R.: Virtual Reality Educational Tool for Human Anatomy. J. Med. Syst. **41**, 76 (2017). https://doi.org/10.1007/s10916-017-0723-6
10. Alpaugh, K., Ast, M.P., Haas, S.B.: Immersive technologies for total knee arthroplasty surgical education. Arch. Orthop. Trauma. Surg. **141**, 2331–2335 (2021). https://doi.org/10.1007/s00402-021-04174-7
11. Hettig, J., Engelhardt, S., Hansen, C., et al.: AR in VR: assessing surgical augmented reality visualizations in a steerable virtual reality environment. Int. J. Cars. **13**, 1717–1725 (2018). https://doi.org/10.1007/s11548-018-1825-4
12. Huang, Y.C., Backman, S.J., Backman, K.F., et al.: An investigation of motivation and experience in virtual learning environments: a self-determination theory. Educ. Inf. Tech. **24**, 591–611 (2019). https://doi.org/10.1007/s10639-018-9784-5
13. Pasfield-Neofitou, S., Huang, H. and Grant, S.: Lost in second life: virtual embodiment and language learning via multimodal communication. Educ. Tech. Res. Dev. **63**, 709–726 (2015). https://doi.org/10.1007/s11423-015-9384-7
14. Doğan, D., Demir, Ö., Tüzün, H.: Exploring the role of situational flow experience in learning through design in 3D multi-user virtual environments. Int. J. Tech. Des. Educ. **32**, 2217–2237 (2022). https://doi.org/10.1007/s10798-021-09680-8
15. Elme, L., Jørgensen, M.L.M., Dandanell, G., et al.: Immersive virtual reality in STEM: is IVR an effective learning medium and does adding self-explanation after a lesson improve learning outcomes?. Educ. Tech. Res. Dev. (2022). https://doi.org/10.1007/s11423-022-10139-3
16. Evans, S.: Virtual Selves, Real Relationships: An Exploration of the Context and Role for Social Interactions in the Emergence of Self in Virtual Environments. Integr. Psych. Behav. **46**, 512–528 (2012). https://doi.org/10.1007/s12124-012-9215-x
17. Zimmermann, D., Wehler, A., Kaspar, K.: Self-representation through avatars in digital environments. Curr. Psych. (2022). https://doi.org/10.1007/s12144-022-03232-6
18. Hammady, R., Ma, M., Strathern, C., et al.: Design and development of a spatial mixed reality touring guide to the Egyptian museum. Multi. Tools. Appl. **79**, 3465–3494 (2020). https://doi.org/10.1007/s11042-019-08026-w
19. Torres-Ruiz M., Mata, F., et al.: A recommender system to generate museum itineraries applying augmented reality and social-sensor mining techniques. Virtual. Reality. **24**, 175–189 (2020). https://doi.org/10.1007/s10055-018-0366-z
20. Johnson, D.G.: Ethics Online. Communications of the ACM. **1**, 60–65 (1997). https://doi.org/10.1145/242857.242875
21. Floridi, L.: A Look into the Future Impact of ICT on Our Lives. Info. Soc. **23**, 59–64 (2007). https://doi.org/10.1080/01972240601059094
22. Baraniuk, C.: Police to test app that assesses suspects. BBC News. (2017). Retrieved from http://www.bbc.com/news/technology-39857645
23. The Economist.: Don't even think about it. The Economist. (2013). Retrieved from https://www.economist.com/news/briefing/21582042-it-getting-easier-foresee-wrongdoing-and-spot-likely-wrongdoers-dont-even-think-about-it

24. Moore, A.D.: Privacy: Its meaning and value. Am. Philos. Quar. **40**(3), 215–227 (2003)
25. Allen, A.: Privacy and medicine. In: Zalta, E.N. (Ed.) The Stanford Encyclopedia of Philosophy. (2011). Retrieved from http://plato.stanford.edu/archives/spr2011/entries/privacy-medicine/
26. Yee, N.: The psychology of massively multi-user online role-playing games: Motivations, emotional investment, relationships and problematic usage. In: Avatars at work and play, 187–207 (2019). Netherlands: Springer
27. Blackburn, T.: Chinese Gamer Dies After 40 Hour MMO Session. (2014). http://www.gamebreaker.tv/game-industry-news/chinese-gamer-dies-after-40-hour-mmo-session/

The Use of Metaverse in the Healthcare Sector: Analysis and Applications

Rania. A. Mohamed, Kamel K. Mohammed, Ashraf Darwish, and Aboul Ella Hassanien

Abstract With the use of blockchain, Internet of Things, virtual platform/ telecommunications network, artificial intelligence and the fourth industrial revolution, the essential demand for digital transition within the health care settings has increased as an outcome of the 2019 coronavirus illness outbreak and the fourth industrial revolution. The evolution of virtual environments with three-dimensional (3D) spaces and avatars, known as metaverse, has slowly gained acceptance in the field of health care. These environments may be especially useful for patient-facing platforms (such as platforms for telemedicine), functional uses (such as meeting management), digital education (such as modeled medical and surgical learning), treatments and diagnoses. This chapter offers the most recent state-of-the-art metaverse services and applications and a growing problem when it comes to using it in the healthcare sector.

1 Introduction

A significant and brand-new type of digital civilization is emerging due to the COVID-19 pandemic and the fourth industrial era. A company is the most significant element in a contemporary civilization that links any person and society. The metaverse is the subject of technical inventions, boosting business opportunities in the new digital civilization. Due to the COVID-19 incident, the business environment

R. A. Mohamed (✉)
Faculty of Computers and Artificial Intelligence, Modern University for Technology and Information, Cairo, Egypt
e-mail: rmohamed@cs.mti.edu.eg

K. K. Mohammed
Center for Virus Research and Studies, Al-Azhar University, Cairo, Egypt

A. Darwish
Faculty of Science, Helwan University, Cairo, Egypt

A. E. Hassanien
Faculty of Computer and Artificial Intelligence, Cairo University, Cairo, Egypt

A. E. Hassanien et al. (eds.), *The Future of Metaverse in the Virtual Era and Physical World*, Studies in Big Data 123, https://doi.org/10.1007/978-3-031-29132-6_5

around the metaverse has significantly altered. Online business is becoming more prevalent across all company sectors, replacing the previous consumer mentality that focused on face-to-face interactions. The utilization of intelligent technologies in the healthcare sector has become grown significantly. It is obvious that the zero-contact business culture has had a significant impact on the healthcare sector, as the selection of smart services applications has grown and the demand for smart services has improved greatly. The healthcare sector is becoming more and more promising and important as a result of the applications of metaverse services [1–5]. Consequently, it is crucial that the healthcare sector understands metaverse services as a crucial component of operational planning for the sector's long-term vision. The idea of the metaverse has been actively debated since 2021. As the prospective mobile computing platform that will have significant future usage, it depicts the internet that is accessible with VR and AR eyewear. Others contend that the metaverse is a ternary digital space that combines the virtual and actual worlds and that people enter using their online profiles. The book True Names by American mathematician Professor Vernor Vinge served as the inspiration for the concept. The author of this 1981 novella imaginatively imagined a virtual world that might be accessed and sensed via a brain-computer link. Later, in 1992, American science-fiction author Neal Stephenson used the phrase "metaverse" in his book Snow Crash, in which the protagonists explore a parallel internet universe to the actual one while interacting with digital avatars of themselves. Possibilities for all sectors of society and professions have increased as a result of the metaverse's growth, including the creation of video games and other forms of leisure and entertainment. Diverse digital technologies have helped museums update their exhibits, while conventional shopping has given rise to new metaverse sales models [6]. Investigators have looked at the 3D virtual world's artistic community. Mark Zuckerberg announced that Facebook has renamed itself to "Meta" in order to better represent the company's emphasis on cutting-edge computing innovations and the metaverse on October 28, 2021. Others are interested in learning about how journalism is carried out in the metaverse. It was recently recommended that 2022 be designated as the Year of the Metaverse in Medicine by Bai and colleagues [7]. The expert panel also examined the metaverse's notion in relation to medicine, as well as its theoretical underpinnings, potential applications, and clinical significance. The goal of this new concept is to transform the current model of identification and treatment, this varies across physicians and medical facilities and leads to inconsistent standards, akin to those produced in handicraft factories (known as the "handicraft workshop concept " herein), into a contemporary assembly-line model that complies with national and even global benchmarks.

This chapter is organized as following. Section 2 presents the importance of the Metaverse. Section 3 highlight the concepts of the medical Metaverse. In section 4, the application of the Metaverse in the medical sector are presented. Section 5 presents the problems, challenges and future trends. Section 6 concludes this chapter.

2 Importance of the Metaverse

In industrialized nations, economic expansion and ongoing improvements in living standards have coexisted with population aging, lifestyle changes, and a variety of illnesses, creating new challenges for healthcare. Participating in health governance and carrying out national commitments to the UN 2030 Agenda for Sustain Development are equally crucial in those nations. Building a healthy nation calls for increased efforts in encouraging scientific and technical innovation, enhancing organization and implementation, optimizing healthcare service systems, and creating digital health information services. A recent description of the Metaverse in Medicine, put out by Professor Chunxue Bai and colleagues, describes it as the medical Internet of Things (MIoT) made possible by AR and/or VR glasses. The development of the MIoT has provided a timely resolution to these issues since it provides excellent technical support. According to the present MIoT theory, it is possible to achieve effective and precise graded detection and therapy by connecting doctors from large hospitals (referred to as "Cloud Experts") with doctors from smaller hospitals (referred to as "Terminal Doctors"), as well as to participate to the research and advancement of related technologies to enhance graded assessment and therapy. China, for instance, has issues with regional and hospital health resource inequities. Small rural hospitals frequently lack access to cutting-edge medical equipment (referred to as "inadequate device coverage"), local physicians frequently lack technical expertise (referred to as "inadequate technical competence"), and patients frequently express low levels of satisfaction with their care (referred to as "inadequate patient satisfaction"). Many people choose to visit major hospitals and see renowned physicians for better treatment and diagnosis as a consequence of these "Three Deficiencies," which leads to registration and hospitalization issues that are known as the "Two Difficulties." [8–11]. Moreover, as a result of the inflow of rural patients into city hospitals which creates each expert is able to spend a limited amount of time with each patient, resulting in the distribution of services for disease prevention, treatment, healthcare, and rehabilitation, which we denote by the phrase "Four Limitations". We advised using the MIoT to help physicians in clinical practice to solve these challenges, and specifically suggested leveraging the three core MIoT capabilities of comprehensive observation, reliable transmission, and intelligent process, which have been effectively used in many situations [8–10]. Additionally, it is expensive to provide people with quality and easily accessible healthcare services owing to the rising need for medical services and the huge number of professionals engaged. The Internet of Things (IoT), which combines telecommunications with smart mobile devices, has the potential to significantly address this issue. One of the most widely utilized breakthroughs in the e-health sector has been the MIoT, which has moved medical services from hospitals to the workplace and homes [12]. The generation of knowledge in the MIoT and related domains has greatly expanded since 2018. Additionally, the COVID-19 pandemic has raised awareness of the need for home medical services, which is one of the goals of e-health, and more specifically of the MIoT. On the one hand, IoT programs are often created to save expenses, provide patients increased accessibility at home, and promote patient participation, all of which contribute to increasing healthcare and people's welfare. On the other hand, Grieves presented the Digital Twin concept as a new industry standard 4.0 in 2002. [13] enables the integration of VR and AR technologies into the MIoT and accelerates the transition into highly effective clinical applications (The IEEE Digital Twin: Enabling Technologies, Difficulties, and Open Investigation). But being a growing discipline in applied research, the MIoT confronts a variety of difficulties, much like any other new medical technology. These concerns need to be confirmed and resolved by extensive clinical application and marketing in the near future. They include challenges with medical supervision, medical insurance, and the digital splitting [14]. According to Bai et al., the MIoT will develop into a school of thought [7] and a useful medical tool because it can "simplify difficult concerns, digitalize simple issues, program digital challenges, and systematize programming problems." The ultimate objective is to transform China's healthcare and medicine from the existing model, which differs across hospitals and physicians and has unequal service rates, to a cutting-edge assembly-line model that satisfies national and even worldwide standards, realizing our vision: Great physicians help the general population by treating people before illnesses develop.

3 The Medical Metaverse

There will be four phases in the growth of the clinical metaverse (meta-medicine) (Fig. 1).

1. **Holographic construction**

Constructing a static geometric model of the complete virtual environment, including medical centers, equipment, and other items, will be the initial step. These items may be broken down into three primary categories: events, scenes, and individuals. For instance, in a normal surgical scenario, the interior setting and medical gear make up the "scene," while the patients and medical personnel make up the "people." Events are made up of dynamic variables produced by interactions between individuals or between individuals and settings.

2. **Simulation with holograms**

The dynamic meta-medical procedure will be designed to create a virtual world as close to the real world as possible. On the one hand, technology will advance alongside us, enhancing fidelity at the physical level, coarse- and fine-grained threshold, and communication degree. It will be connected to PACS systems and other medical information systems on the opposite side. Digital patient and physician avatars will provide quick, logical feedback on pertinent knowledge and data via the use of actual-time movement capture and multi-sensor devices for informative engagement.

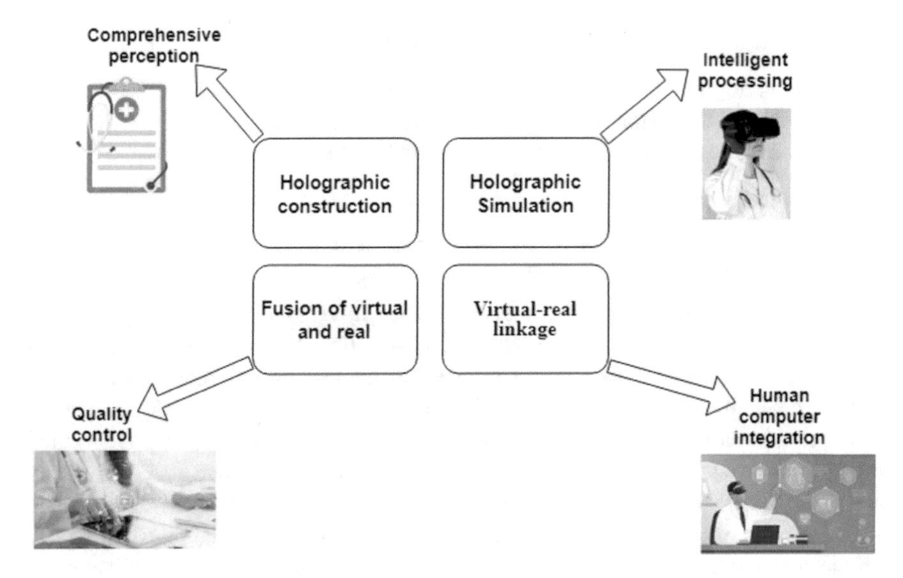

Fig. 1 Four phases of the medical metaverse

3. **Virtual and real fusion**

The virtual medical environment will seem more lifelike because to digital technology's ongoing advancements in computer simulation technologies. The line between the virtual and actual worlds will be broken by creating a mixed reality (MR) environment. A doctor, for instance, may speak with a patient using an XR device in an outpatient environment. Patient-related data will be gathered in real-time and displayed as an overlaid reality via real-time recognition, tracking, and data engagement, creating a new connection with the surrounding natural environment.

4. **Virtual-real linkage**

The ongoing advancement of simulation technologies, brain-computer interfaces (BCI), artificial intelligence (AI) [15, 16], and other techniques will lead to the creation of many new medical procedures, kinds of medical equipment, and facilities. Real-world modifications to medical procedures will follow advances in technology. In the age of meta-medicine, medical treatments on the two planets will progressively create more intersections and eventually give rise to new ideas and techniques. The meta-medical community will support and advance current medical procedures in the actual world. Through the information space, we will be able to not only observe but even alter the physical world. Three main qualities will set meta-medicine apart from conventional medicine.

(1) Interactions of all things

Everything in the actual world is linked to the virtual world, allowing patients and physicians to view and alter it in real-time using a variety of technologies.

(2) The integration of virtual and actual objects

Through economic, social, and identification systems, meta-medicine will tightly link the real and the virtual so that patients and physicians may experience the same richer, more intuitive awareness in the 3D world of the actual and the virtual.

(3) Decentralization

Blockchain technology has the potential to be used for virtual identity distribution, trade, usage, liquidity, and digital medical resources that need authentication as well as for the intelligent administration of digital rights [17]. The 3D elements of the metaverse are users, technology, and information. Patients and physicians are the two most significant user groups in meta-medicine. Through digital twins, each person in the virtual world will have their own "digital identities." The digital avatars of the meta-medical world will take on the characteristics of current patient data, imaging information, and other attributes. Through their respective avatars, doctors and patients will be able to converse without being constrained by time or geography. Based on this, the digital world will have more flexibility than the physical world, providing us the ability to observe things in more dimensions, communicate and interact differently, and modify "reality" in the context of meta-medicine. A trustworthy technological system that is based on Web3.0, a "distributed" and "traversing" Internet that is created in five tiers, will be necessary for the implementation of meta-medicine [17]. The foundation of a network connection is made up of the IoT and a 5G network environment. For instance, China has gathered significant expertise in IoT medical research on respiratory disorders, some of which has been utilized in clinical practice. Early identification and treatment of asthma, chronic obstructive pulmonary illness, lung carcinoma, and other persistent conditions have all benefited from AI. It also features sophisticated auxiliary viral pneumonia diagnosis and therapy, among other things [18]. The basis for computing power at the level of data processing has been solidified by edge computing and cloud computing. The cloud is used for mass intelligent analysis of medical data under the cloud computing framework, which upholds the "cloud endpoint framework" and the deep processing of huge data. The creation, verification, and exchange of the metaverse's digital resources will operate in an orderly fashion thanks to the technological basis provided by the non-fungible token (NFT) system and blockchain technology at the authentication level [19].

4 Applications of the Medical Sector

4.1 *Metaverse in Ophthalmologic*

The COVID-19 epidemic has had a tremendous influence on people's way of life all around the globe. In order to minimize face-to-face consultations, prioritize cases needing immediate medical attention, reschedule non-urgent visits, and implement new infection control strategies, healthcare organizations had to rearrange treatment for their present patients [20]. To increase the efficiency of the medical practices, lessen the amount of time spend in facilities, and decentralize treatment with fewer physical touchpoints, there is a need for and potential for digital transformation [21]. The process of transforming an entity via the use of information, computer, communication, and connection technologies is known as digital transformation [22]. For instance, up to 77.5% of Indian ophthalmologists employed teleophthalmology to offer eye care during the COVID-19-induced lockout in 2020 [23], while American neuro-ophthalmologists expected a 64.4% rise in the use of telehealth procedures [24]. Ophthalmologists in China revealed that 10,000 visits involving the eyes were handled by an AI chatbot and that they employed digital tools for virtual consultations [25]. Particularly during the COVID-19 epidemic or in rural areas with few ophthalmology care, synchronous telemedicine has shown to be effective for emergency department consults in triaging the urgency of referrals or giving ocular first aid remotely. For instance, during the COVID-19 epidemic, Moorfields Eye Hospital in London started offering emergency video consultations to reduce the number of ocular casualties on its physical premises. It was shown that this service prevented more than 70% of possible in-person interactions [26]. Similar to this, emergency ophthalmic teleconsultation was used in Paris during the COVID-19 shutdown in 2020, which resulted in a 73% reduction in pointless physical consultations [27]. Outpatient ophthalmology consultations may also be conducted by real-time video. A tertiary eye care facility in Singapore found increased sensitivity and specificity when employing this approach to assess chronically impaired vision [28]. General ophthalmic facilities in Hong Kong were able to retain 80% of their outpatient load during the outbreak by utilizing an input funnel model [29]. To conduct a teleconsultation, the clinician would get clinical information and images from the distant providers (an optometrist or ophthalmic specialist). With positive feedback from physicians and patients, subspecialties such oculoplastic [30], pediatric ophthalmology, and strabismus [31] have also used video-based teleconsultations. Additionally, studies have shown that the majority of non-urgent cases may be handled exclusively by teleconsultation or can be used as a useful triaging tool for a hospital evaluation. In addition, owing to the convenience and safety offered by online platforms, 62% of patients with oculoplastic disorders chose a video consultation over a face-to-face appointment. Virtual consultations thus have the potential to be extensively used as an alternate model of treatment to provide extremely reliable diagnosis eye care that is backed by public approval, even when regular health facilities restart. In addition to video meetings, synchronous telemedicine may be used for "telemonitoring," [32] which is the delivery of in-procedure instructions and direction to a surgeon by a specialist located elsewhere in the world. Telemonitoring has the ability to provide generalists and trainees with access to specialized training on par with on-site instruction [33]. There aren't many instances available right now to show how successful telemonitoring in ophthalmology is. A general ophthalmologist safely removed an orbital tumor with help from a 210-mile-away [34] orbital specialists through telementoring, according to a case report from 2000. Another instance demonstrated the transfer of knowledge from ophthalmologists to general practice in a rural hospital to assist the removing a foreign object from the cornea while pictures from the slit-lamp were being sent in real-time [35]. More recently, two vitreoretinal surgical cases that were successfully teleproctored via a 5G network employing real-time three-dimensional (3D) video transmission were described [36]. Telemonitoring can thereby overcome the variation in access to healthcare on a global and national scale. On the other hand, telesurgery occurs when a surgeon operates on a patient from a distance [37]. Because it is synchronous, a solid network with low latency and high data transmission speeds is needed (e.g., 5G network) [38]. The majority of telesurgery cases to far have been performed on animals or via simulation. In two ophthalmological experiments, telerobotic surgery was used to performing transscleral cyclophotocoagulation on human eyes with enucleation [39] and to treating rabbit eye corneal abrasions [40]. Separately, this field has had early success for other surgical subspecialties including orthopedics [41] and urology [42]. For instance, a Chinese orthopedics team used a 5G network technology to effectively execute telerobotic spinal surgery across several locations with no intraoperative negative events. Therefore, the capability of remote surgery, which crosses geographic barriers, provides the potential to transform the provision of healthcare and access to specialized treatment. But telesurgery is still in its early stages, and more study and development are needed before it can be used in clinical settings. It should be remembered that the price of the necessary equipment may still act as a deterrent to adoption.

4.2 Application of the Metaverse in Gastrointestinal

Due to time and space constraints, many gastrointestinal endoscopists now, particularly in light of the pandemic, are unable to go to wealthy nations for long-term professional teaching and acquire knowledge of the most recent endoscopic technique [43]. However, with the aid of VR, AR, and other technologies, we may take on the form of virtual beings in the Metaverse. By viewing a live broadcast of an endoscopy expert's procedure on the other side of the globe, we may learn the expert's procedure on the spot rather than just looking at the endoscopic screen. A Metaverse lung cancer training exercise took place on May 29, 2021, at the computerized lecture hall of Seoul National University in South Korea. Up to 200 Asian thoracic specialists were able to view the Metaverse via a head-mounted display. After finishing their personalized settings, participants went to a virtual classroom where they saw a 360-degree teaching course for lung tumor surgeries and a presentation on metasomatic technological developments. While viewing surgical training, experts converse with one another in real time via their avatars and debate techniques in the Metaverse. This is the first account of the metaverse's usage in surgical training anywhere in the globe, despite the fact that it is exceedingly small [44]. Additionally, gastrointestinal endoscopists may search the extensive internet library for the fundamental subject matter and training programs necessary in the Metaverse. Advanced endoscopic technology allows them to experiment on virtual patients in the Metaverse. Senior physicians may provide remote counseling and treatment for these patients, enabling high-quality healthcare facilities to help more people and, to some part, resolving the issues connected with the current unequal dispersion of health resources. In the many less developed countries and areas of the globe, where there is still a significant lack of digestive doctors and endoscopic technology, patients may also be treated by senior physicians. The underlying technologies of the Metaverse, including as virtual reality (VR), haptic gloves, robots that operate remotely, and wireless technology for communication have progressed quickly and are used in clinical evaluation and therapy, therefore these ideas are obviously not just science fiction. We might investigate future applications of Metaverse in gastroenterology by introducing the applying of these fundamental Metaverse methodologies. Since at least 20 years ago, VR, one of the meta-subtechnologies, has been widely employed in endoscopic professional training [45]. Endoscopic trainee doctors were trained using GI Mentor, simulators for virtual endoscopy's first generation, as early as 2002 [46]. The computer creates a 3D image of the simulated digestive system when the endoscope is introduced into the gastrointestinal model. When the simulated intestinal wall is contacted, the force feedback module will simulate resistance at the same time, giving the user a realistic surgical experience. The improved Pentax ECS-3840F enables users to carry out the treatment, which quickly raises the endoscopic level of beginning doctors. Accutouch, a technique for VR colonoscopy simulation, was created in 2005 to aid experts in performing endoscopic colorectal cancer surgery more effectively [47]. It provides professionals with more practical operational knowledge and a multimedia training and assistance system in contrast to GI Mentor. The colonoscopy training may be completed by beginners by obediently performing computer commands. The Olympus Simulation for Colonoscopy (Endo TS1) was created by advent of Olympus KeyMed, Southend, UK, in 2008 to provide younger endoscopists a more tactile prior training for VR endoscopy [48]. A growing number of potent VR digestive endoscopy simulations have been employed in medical practice as VR technology has advanced in recent years. A new iteration of the GI Mentor II medical simulations education program was released in 2015 [49]. Additionally, it may simulate a variety of colorectal diseases, giving users a more authentic endoscopic experience. VR endoscopy training has been shown in a growing number of clinical trials to dramatically increase physicians' expertise, raise the caliber of endoscopic procedures, cut down on procedure time, and lower the frequency of surgical consequences [50–53]. Simulator-based training improves several interventional abilities, including endoscopic haemostasis, perforation closure, and retrograde cholangiography, according to investigations of the Erlangen Endo trainer [54]. However, these models do have some drawbacks. For instance, the VR model's illnesses are restricted to routine exams, endoscopic haemostasis, and closed perforation, making them poor training tools for difficult procedures like endoscopic mucosal resection (EMR), endoscopic submucosal dissecting (ESD), and endoscopic ultrasound (EUS) [55]. Additionally, although such training is advantageous for starting endoscopists, it does not improve the proficiency of experienced endoscopists [56, 57]. The growth of endoscopists is greatly constrained by the fact that the current generation of endoscopic simulators only offers endoscopists fundamental teaching and cannot accurately duplicate more sophisticated surgical procedures like EMR, ESD, or other therapeutic techniques

[58]. However, since the Metaverse is a virtual environment that is wholly distinct from the actual world, it will ultimately be able to provide us with all the situations we need to imitate, enabling us to get over the limitations of models and technology. Only those who have completed extensive endoscopic training and have access to Metaverse equipment, VR helmets, and tactile gloves, should use them. Moreover, we could utilize AI to look for cloud databases for expert instruction videos and develop the skills necessary to operate close-up endoscopes in Metaverse [59]. As a consequence, we not only obtain vast endoscopic competence but also reduce the unnecessary risk of surgical failure.

4.3 Use of the Metaverse in Prenatal and Gynecological Medicine

Healthcare has already benefited from AR resources and will continue to benefit greatly from this technology advancement [60–62]. We have already begun using the metaverse in gynecological and fetal medicine. Our initial goal is to teach experts from many fields in interdisciplinary talks that found in actual time and are not geographically constrained. We do this by including features like the capacity to review a succession of 2D slice imagery, as well as shared navigation within a 3D organ. A remote class in which the teacher presents a case with students participating in VR from various places is an illustration of this situation. They watch the results of a prenatal magnetic resonance examination and interact with it to move through 3D data slices [63]. Additionally, they look at a 3D virtual model that has the organs of the fetus segmented. Another illustration demonstrates the potential for a virtual conference to take place within the 3D layout of a fallopian tube that was high resolution micro-CT scanning [63]. These approaches were created using the versatile Spatial.io platform [64], which accommodates a variety of gadgets. In the upcoming years, we want to assess the usage of this technology while also incorporating the patient in a multidisciplinary team discussion. A multidisciplinary team and students may benefit greatly from using Metaverse as a digital tool to better collaborate and comprehend complicated prenatal abnormalities and gynecological diseases.

4.4 Use of the Metaverse in Lung Cancer

For instance, in 2014 the Zhongshan Hospital of Fudan University and the Chinese Alliance Against Lung Cancer made the initial identification of a pulmonary nodule and treating sub-center. The terminal works closely with a variety of experts from the pulmonary, thoracic, and radiological departments at the respiratory nodule special outpatient clinic. Patients with malignant and benign lung nodules were accurately diagnosed by Professor Bai's team using IoT medicine. A sum of 16,400 lung nodule procedures were carried out in the six years leading up to 2019, and pathological analysis revealed that 60.8% of these procedures were for respiratory tumor that was in the initial stages. These patients with early-phase pulmonary cancer provide great findings for the nation, the patients, and the healthcare community since they do not need postoperative chemotherapeutic, radiotherapy, targeted treatment, or immunotherapy. It will hasten the achievement of the Chinese Alliance Against Lung Cancer's smart therapy of millions of initial tumor targets if it can be applied to the 900 lung nodules identification and treating sub-centers developed by the Metaverse in Medical Fusion-Quality Control Technology [65].

5 Problems, Challenges and Future Perspectives

The capacity for individuals and organizations to build their own virtual worlds is something platform providers are working to improve, but there are several hurdles from a sociotechnical and governance viewpoint. Substantial areas of concern about ethics, data protection, legislation, safety, as well as the possible negative psychological effects on vulnerable people of society, have been uncovered via studies. Users are claiming an increase in rude and undesired behaviors in the metaverse's current spaces, such as user harassment, Institutional mechanisms are required for concerns such drug surrogate therapy [66], the cautious handling of medical information (such as user authorization and privacy), and data exploitation.

6 Conclusion

The growth of digital or precision medicine will be aided by the metaverse's rise, which will open up endless possibilities in the field of healthcare. The development of digital innovations, including 5G information, digital twinning, AI, VR, and blockchain technology, provides the technical foundation for creating the metaverse. The presentation of intricate medical information and data will be intuitive. In an immersive setting, high-cost medical cooperation, experimentation, education, and digital avatars will be used to remotely manage other operations. In order to create, adapt, and share knowledge in the metaverse, patients and physicians will employ "digital avatars," which will allow them to eventually enter the world of virtual-real integration and connection. Furthermore, the ongoing advancement of digital technology unavoidably comes with certain unnoticed risks. The medical industry is distinct from other industries and has its own set of ethics and regulations. Understanding the limits of meta-medicine, protecting the personal data and privacy of patients, and creating suitable laws and regulations for managing systems are still important issues to investigate.

The metaverse has the potential to transform the way healthcare is delivered and experienced in medicine. It can be used to build virtual reality (VR) simulations for medical training, for example, allowing healthcare practitioners to practice operations and techniques in a safe and controlled setting. It can also be used for remote consultations and telemedicine, allowing patients to get medical care from anywhere in the globe.

Furthermore, the metaverse can be utilized to construct virtual support groups and communities for people suffering from chronic diseases, offering a sense of social support and connection that can be beneficial to their mental and physical health.

Overall, the metaverse has the potential to significantly improve healthcare delivery and patient experience. It is an intriguing and quickly evolving field that is likely to have significant future implications in the realm of medicine.

References

1. Holloway, D.: Virtual worlds and health: healthcare delivery and simulation opportunities. In: Virtual Worlds and Metaverse Platforms: New Communication and Identity Paradigms, pp. 251–270. IGI Global: Hershey, PA, USA (2012)
2. Seo, K.R., Park, S.H.: Opportunities and Challenges of the Metaverse Government. National Information Society Agency, Daegu, Korea (2021)
3. Ghanbarzadeh, R., Ghapanchi, A.H., Blumenstein, M., Talaei-Khoei, A.: A decade of research on the use of three-dimensional virtual worlds in health care: a systematic literature review. J. Med. Internet Res.**16**, e3097 (2014). [CrossRef] [PubMed]
4. Kovacev, N.: Metaverse and medicine. In: Proceedings of the 2022 IEEE Zooming Innovation in Consumer Technologies Conference (ZINC), Novi Sad, Serbia, 25–26 May 2022; p. 1
5. Thomason, J.: Metahealth-How will the metaverse change health care? J. Metaverse **1**, 13–16 (2021)
6. Bourlakis, M.: Li PFJECR. Retail spatial evolution: paving the way from traditional to metaverse retailing (2009)
7. Yang, D.W., Zhou, J., Song, Y.L., Bai, C.X.: Metaverse in medicine. Clinical eHealth; epub ahead of print
8. Bai, C.X.: Practical Medical Internet of Things. People's Medical Publishing House, Beijing (2014)
9. Bai, C.X.: Guidelines on Applying Medical Internet of Things for the Graded Diagnosis and Treatment. People's Medical Publishing House, Beijing (2015)
10. Bai, C.X., Zhao, J.L.: Medical Internet of Things. Science Press, Beijing (2016)
11. Bai, C.: Letter from China. Respirology **23**, 718–719 (2018)
12. Su, X., Bai, C.X.: Leveraging cloud computing and terminal to embrace the new era of medical internet of things. China Med Pharm. **6**, 1–3 (2016)
13. Grieves, M.: Virtually intelligent product systems: digital and physical Twins. In: Complex Systems Engineering. Theory and Practice (2019)
14. Sim, I.: Mobile devices and health. N Engl J Med. **381**, 956–968 (2019)
15. Darwish, A., Sarkar, M., Panerjee, S., Elhoseny, M.: Aboul Ella Hassanien, exploring new vista of intelligent collaborative filtering: a restaurant recommendation paradigm. J. Comput. Sci. **27**, 168–182 (2018)
16. Eid, H.F., Darwish, A., Hassanien, A.E., Kim, T.: Intelligent hybrid anomaly network intrusion detection system. In: Communication and Networking. FGCN 2011. Communications in Computer and Information Science, vol. 265. FGCN 2011, Part I, CCIS 265, pp. 209–218. Springer, Berlin, Heidelberg (2011)
17. Wu, J.C.Z., Chen, P., He, C.C., Ke, D.: User information behavior from the metaverse perspective: framework and prospects. J. Inform. Resour. Manag., 1–17

18. Dawei Yang, C.B.: Development status and trend of Internet of things in the medical field. China Med. Inform. Herald. **36**(19), 1 (2021)
19. Nadini, M., Alessandretti, L., Giacinto, F.D., Martino, M., Aiello, L.M., Baronchelli, A.: Mapping the NFT revolution: market trends, trade networks and visual features. Papers (2021)
20. Gunasekeran, D.V., Tham, Y.C., Ting, D.S., et al.: Digital health during COVID-19: lessons from operationalising new models of care in ophthalmology. Lancet Digit Health. **3**, e124–e134 (2021)
21. Li, J.P.O., Shantha, J., Wong, T.Y., et al.: Preparedness among ophthalmologists: during and beyond the COVID-19 pandemic. Ophthalmology **127**, 569–572 (2020)
22. Vial, G.: Understanding digital transformation: a review and a research agenda. J. Strateg. Inf. Syst. **28**, 118–144 (2019)
23. Nair, A.G., Gandhi, R.A., Natarajan, S.: Effect of COVID-19 related lockdown on ophthalmic practice and patient care in India: results of a survey. Indian J. Ophthalmol. **68**, 725–730 (2020)
24. Moss, H.E., Lai, K.E., Ko, M.W.: Survey of telehealth adoption by neuro-ophthalmologists during the COVID-19 pandemic: benefits, barriers, and utility. J. Neuroophthalmol. **40**, 346–355 (2020)
25. Wu, X., Chen, J., Yun, D., et al.: Effectiveness of an ophthalmic hospital based virtual service during the COVID-19 pandemic. Ophthalmology **128**, 942–945 (2021)
26. Kilduff, C.L., Thomas, A.A., Dugdill, J., et al.: Creating the Moorfields' virtual eye casualty: video consultations to provide emergency teleophthalmology care during and beyond the COVID-19 pandemic. BMJ Health Care Inform. **27**, e100179 (2020)
27. Bourdon, H., Jaillant, R., Ballino, A., et al.: Teleconsultation in primary ophthalmic emergencies during the COVID-19 lockdown in Paris: experience with 500 patients in March and April 2020. J. Fr. Ophtalmol. **43**, 577–585 (2020)
28. Tan, J.C., Poh, E.W., Srinivasan, S., et al.: A pilot trial of tele-ophthalmology for diagnosis of chronic blurred vision. J. Telemed. Telecare **19**, 65–69 (2013)
29. Wong, J.K.W., Shih, K.C., Chan, J.C.H., et al.: Tele-ophthalmology amid COVID-19 pandemic-Hong Kong experience. Graefes Arch. Clin. Exp. Ophthalmol. **259**, 1663 (2021)
30. Kang, S., Thomas, P.B.M., Sim, D.A., et al.: Oculoplastic video-based telemedicine consultations: Covid-19 and beyond. Eye (Lond). **34**, 1193–1195 (2020)
31. Deshmukh, A.V., Badakere, A., Sheth, J., et al.: Pivoting to teleconsultation for pediatric ophthalmology and strabismus: our experience during COVID-19 times. Indian J. Ophthalmol. **68**, 1387 (2020)
32. Huang, E.Y., Knight, S., Guetter, C.R., et al.: Telemedicine and telementoring in the surgical specialties: a narrative review. Am. J. Surg. **218**, 760–766 (2019)
33. Erridge, S., Yeung, D.K.T., Patel, H.R.H., et al.: Telementoring of surgeons: a systematic review. Surg. Innov. **26**, 95–111 (2019)
34. Camara, J.G., Zabala, R.R., Henson, R.D., et al.: Teleophthalmology: the use of real-time telementoring to remove an orbital tumor. Ophthalmology **107**, 1468–1471 (2000)
35. Hall, G., Hennessy, M., Barton, J., et al.: Teleophthalmology-assisted corneal foreign body removal in a rural hospital. Telemed. J. E Health **11**, 79–83 (2005)
36. Lu, E.S., Reppucci, V.S., Houston, S.K.S., et al.: Three-dimensional telesurgery and remote proctoring over a 5G network. Digit. J. Ophthalmol. **27**, 38–43 (2021)
37. Hung, A.J., Chen, J., Shah, A., et al.: Telementoring and telesurgery for minimally invasive procedures. J. Urol. **199**, 355–369 (2018)
38. Singh, G., Casson, R., Chan, W.: The potential impact of 5G telecommunication technology on ophthalmology. Eye (Lond). **35**, 1859–1868 (2021)
39. Belyea, D.A., Mines, M.J., Yao, W.J., et al.: Telerobotic contact transscleral cyclophotocoagulation of the ciliary body with the diode laser. J. Robot Surg. **8**, 49–55 (2014)
40. Mines, M.J., Bower, K.S., Nelson, B., et al.: Feasibility of telerobotic microsurgical repair of corneal lacerations in an animal eye model. J. Telemed. Telecare. **13**, 95–99 (2007)
41. Tian, W., Fan, M., Zeng, C., et al.: Telerobotic spinal surgery based on 5G network: the first 12 cases. Neurospine. **17**, 114–120 (2020)

42. Bove, P., Stoianovici, D., Micali, S., et al.: Is telesurgery a new reality? Our experience with laparoscopic and percutaneous procedures. J. Endourol. **17**, 137–142 (2003)
43. Waschke, K.A., Coyle, W.: Advances and challenges in endoscopic training. Gastroenterology **154**, 1985–1992 (2018). https://doi.org/10.1053/j.gastro.2017.11.293
44. Koo, H.: Training in lung cancer surgery through the metaverse, including extended reality, in the smart operating room of Seoul National University Bundang Hospital. Korea. J. Educ. Eval. Health Prof. **18**, 33 (2021). https://doi.org/10.3352/jeehp.2021.18.33
45. Skibba, R.: Virtual reality comes of age. Nature **553**, 402–403 (2018). https://doi.org/10.1038/d41586-018-00894-w
46. Ferlitsch, A., Glauninger, P., Gupper, A., Schillinger, M., Haefner, M., Gangl, A., et al.: Evaluation of a virtual endoscopy simulator for training in gastrointestinal endoscopy. Endoscopy **34**, 698–702 (2002). https://doi.org/10.1055/s-2002-33456
47. Ahlberg, G., Hultcrantz, R., Jaramillo, E., Lindblom, A., Arvidsson, D.: Virtual reality colonoscopy simulation: a compulsory practice for the future colonoscopist? Endoscopy **37**, 1198–1204 (2005). https://doi.org/10.1055/s-2005-921049
48. Koch, A.D., Haringsma, J., Schoon, E.J., de Man, R.A., Kuipers, E.J.: A second-generation virtual reality simulator for colonoscopy: validation and initial experience. Endoscopy **40**, 735–738 (2008). https://doi.org/10.1055/s-2008-1077508
49. Koch, A.D., Ekkelenkamp, V.E., Haringsma, J., Schoon, E.J., de Man, R.A., Kuipers, E.J.: Simulated colonoscopy training leads to improved performance during patient-based assessment. Gastrointest Endosc. **81**, 630–636 (2015). https://doi.org/10.1016/j.gie.2014.09.014
50. Blackburn, S.C., Griffin, S.J.: Role of simulation in training the next generation of endoscopists. World J. Gastrointest Endosc. **6**, 234–239 (2014). https://doi.org/10.4253/wjge.v6.i6.234
51. Kruglikova, I., Grantcharov, T.P., Drewes, A.M., Funch-Jensen, P.: The impact of constructive feedback on training in gastrointestinal endoscopy using highfidelity virtual-reality simulation: a randomised controlled trial. Gut **59**, 181–185 (2010). https://doi.org/10.1136/gut.2009.191825
52. Harpham-Lockyer, L., Laskaratos, F.M., Berlingieri, P., Epstein, O.: Role of virtual reality simulation in endoscopy training. World J. Gastrointest. Endosc. **7**, 1287–1294 (2015). https://doi.org/10.4253/wjge.v7.i18.1287
53. Hashimoto, D.A., Petrusa, E., Phitayakorn, R., Valle, C., Casey, B., Gee, D.: A proficiency-based virtual reality endoscopy curriculum improves performance on the fundamentals of endoscopic surgery examination. Surg. Endosc. **32**, 1397–1404 (2018). https://doi.org/10.1007/s00464-017-5821-5Frontiers
54. Neumann, M., Mayer, G., Ell, C., Felzmann, T., Reingruber, B., Horbach, T., et al.: The Erlangen Endo-Trainer: life-like simulation for diagnostic and interventional endoscopic retrograde cholangiography. Endoscopy **32**, 906–910 (2000). https://doi.org/10.1055/s-2000-8090
55. Grover, S.C., Garg, A., Scaffidi, M.A., Yu, J.J., Plener, I.S., Yong, E., et al.: Impact of a simulation training curriculum on technical and nontechnical skills in colonoscopy: a randomized trial. Gastrointest Endosc. **82**, 1072–1079 (2015). https://doi.org/10.1016/j.gie.2015.04.008
56. Walsh, C.M., Scaffidi, M.A., Khan, R., Arora, A., Gimpaya, N., Lin, P., et al.: non-technical skills curriculum incorporating simulation-based training improves performance in colonoscopy among novice endoscopists: Randomized controlled trial. Dig. Endosc. **32**, 940–948 (2020). https://doi.org/10.1111/den.13623
57. Piskorz, M.M., Wonaga, A., Bortot, L., Linares, M.E., Araya, V., Olmos, J.I., et al.: Impact of a virtual endoscopy training curriculum in novice endoscopists: first experience in Argentina. Digest Dis. Sci. **65**, 3072–3078 (2020). https://doi.org/10.1007/s10620-020-06532-8
58. Khan, R., Plahouras, J., Johnston, B.C., Scaffidi, M.A., Grover, S.C., Walsh, C.M.: Virtual reality simulation training in endoscopy: a Cochrane review and meta-analysis. Endoscopy **51**, 653–664 (2019). https://doi.org/10.1055/a-0894-4400
59. Ward, S.T., Hancox, A., Mohammed, M.A., Ismail, T., Griffiths, E.A., Valori, R., et al.: The learning curve to achieve satisfactory completion rates in upper GI endoscopy: an analysis of a national training database. Gut **66**, 1022–1033 (2017). https://doi.org/10.1136/gutjnl-2015-310443

60. Castro, P.T., Araujo Jr. E., Lopes, J., Ribeiro, G., Werner, H., Placenta accreta: Virtual reality from 3D images of magnetic resonance imaging. J. Clin. Ultrasound, 1–2 (2021)
61. Uruthiralingam, U., Rea, P.M.: Augmented and virtual reality in anatomical education–a systematic review. Adv. Exp. Med. Biol. **1235**, 89 (2020)
62. Werner, H., Lopes Dos Santos, J.R., Ribeiro, G., Belmonte, S.L., Daltro, P., Araujo Jr. E.: Combination of ultrasound, magnetic resonance imaging and virtual reality technologies to generate immersive three-dimensional fetal images. Ultrasound Obstet. Gynecol. **50**(2), 271–272 (2017)
63. Werner, H., Ribeiro, G., Arcoverde, V., Lopes, J., Velho, L.: The use of metaverse in fetal medicine and gynecology. Eur. J. Radiol. **150**, 110241 (2022). https://doi.org/10.1016/j.ejrad.2022.110241
64. Spatial.io, URL: https://spatial.io/
65. Lin Tong, D.Y., Bai, C.: Implications of lung cancer prevention and treatment in the United States for China. Int. J. Respiration. **41**(5), 4 (2021)
66. Anter, A.M., Moemen, Y.S., Darwish, A., Hassanien, A.E.: Multi-target QSAR modelling of chemo-genomic data analysis based on extreme learning machine. J. Knowl. Based Syst., Elsevier Knowl. Based Syst., **188**, 104977 (2020). https://doi.org/10.1016/j.knosys.2019.104977

The Ethics of the Metaverse and Digital, Virtual and Immersive Environments: Metavethics

Matteo Zallio and P. John Clarkson

Abstract While challenges around the creation of safe, ethical digital, virtual environments are as dated as the early attempts to create digital environments in the form of social games, more widespread conversations started only in recent years. With the recent developments in the Augmented Reality and Virtual Reality technology as well as the Metaverse there is a strong need to investigate challenges and opportunities in relation to the creation of safe, inclusive and ethical digital, virtual and immersive environments. With this article we frame a new conversation developed from previous qualitative research around the ethics of the Metaverse, alternatively called, Metavethics. Metavethics refers specifically to the ethics of human behavior towards the Metaverse, as it develops and grows as an advanced, pervasive immersive technology. Metavethics concerns with the study, assessment and safeguard of the ethical, integrity and social implications of the Metaverse, digital environments, virtual spaces and augmented reality content. To thrive, Metavethics requires the combined expertise of specialists from numerous disciplines, who must explore, study and develop critical knowledge to inform the creation of inclusive, safe and secure digital, virtual and immersive environments for all.

Keywords Metaverse · Virtual reality · Augmented reality · Metavethics

1 Introduction

There is no doubt that with the advancements in the design and development of the Metaverse and digital, virtual and immersive environments, more opportunities, as well as challenges, would have matured.

The Metaverse is a set of digital places, including interconnected immersive 3D experiences [1], that enables people to be present in the digital world under various forms.

M. Zallio (✉) · P. J. Clarkson
Engineering Department, Inclusive Design Group, University of Cambridge, Cambridge CB2 1PZ, UK
e-mail: Mz461@cam.ac.uk

A. E. Hassanien et al. (eds.), *The Future of Metaverse in the Virtual Era and Physical World*, Studies in Big Data 123, https://doi.org/10.1007/978-3-031-29132-6_6

With the Metaverse being a digital, virtual and immersive environment where people can be represented by digital copies, named as avatars, there are multiple opportunities to empower people to connect, socialize, and work in a variety of different ways.

The Metaverse, as well as other similar digital, virtual and immersive environments such as the Omniverse [2], or the Multiverse [3], offer limitless opportunities to users to create content that is being used within those digital spaces.

This content of various nature can grow exponentially and be mass distributed across billions of potential users across the world.

However, a series of implications regarding the ethical, social, safety, access and inclusion aspects can develop as potential challenges that could impact the experience of Metaverse users.

Other than the nuances that characterize the development, access, and use of this technology, which are of great importance and fundamental for the creation of seamless experiences, there is a series of other implications that can lead to the creation of safe, secure and inclusive digital, virtual and immersive environments [4].

Some of those implications can be identified with the development of technologies, devices, tools to provide new experiences that are currently not largely available for consumer products [4].

Other implications refer to the design of digital tools, software, token exchange methods and more in general systems that allow an integration between device, data generation and computation and content delivery to the user [4].

Additional implications, which are the current focus of the new discipline of the Metavethics, refer to the intangible nature of digital spaces which generate content that could influence the behavior of people, companies, organizations and more broadly of entire communities.

This content, both digital (e.g., an image, sound, etc.) and physical (e.g., an haptic feedback, a olfactory experience, etc.) which leads to far more immersive experiences than the ones people are currently used to experience with their current devices and apps [5], can lead to the creation of an idea, a change in behavior and could generate a correspondent action in both digital and physical world impacting the security and safety sphere of individuals and communities.

All those implications have strong roots in philosophical, sociological and anthropological aspects that influence the decisions people make. Due to these simple examples, which do not depict the complexity of the entire scenario, it appears that the ethical domain is not deeply explored. Therefore, to increase safety and privacy, improve inclusion and accessibility of digital, virtual and immersive environments and correlated technologies there is a strong need to explore the foundations of the ethical implications of the Metaverse and digital, virtual and immersive environments.

This chapter is organized as follows. Section 2 presents the basics and foundations of the Metaverse. Section 3 highlights the interpretations of ethics. Section 4 presents the ethics of the Metaverse and digital, virtual and immersive environments: the Metavethics. Section 5 concludes this chapter.

2 The Foundations of the Metaverse

One of the first examples of the Metaverse dates back to 1992, the year in which the science fiction novel Snow Crash was launched by Neil Stephenson [6].

In his book, Stephenson mentioned for the first time the term Metaverse, defined as a virtual urban environment that runs around the circumference of a spherical planet.

This book became soon a spark that ignited in the following years and decades the development by technologists, experts in informatics and a series of sci-fi dreamers of virtual spaces in which to test how human beings could live in a digital, parallel world.

One of the most popular examples of a digital, virtual environment was conceived in 2003, when a San Francisco-based company, Linden labs, released Second Life, a virtual space in which people could virtually create a parallel life [7].

Second life was recognized to be one the first attempts to build a digital, virtual environment that lasted for several years [8].

In recent years, large investments from the gaming industry boosted the confidence to create even more impactful and complex digital environments that could be used with virtual reality devices [9].

The gaming and entertainment industry, including companies such as Roblox, Active Worlds, Activision Blizzard, Epic Games, just to name a few, started to invest more in to developing engaging games, supported by technologies that would offer an improved experience than the usual keyboard and screen experience know so far.

Across 2021 and 2022, the hype and popularity around the Metaverse grew exponentially and in 2022 the well-known social media company Facebook updated its brand identity to Meta, as a way to "bring together Facebook apps and technologies under one new company brand and focus on bringing the Metaverse to life by helping people connect, find communities and grow businesses" [10].

To reinforce this recent hype and strong effort on developing digital, virtual environments that provide new experiences to potential users a report from the Gartner research institute reported that by 2026 a quarter of people of the world will be spending at least an hour a day in the Metaverse [11].

Different definitions of the Metaverse appeared, however it seems still difficult to clearly define how far the development of the Metaverse and new immersive, virtual worlds has gone [12].

By looking at trends and recent developments, the Metaverse can be broadly defined as a place of non-places, in which the digital creativity of human brain meets Machine Learning and Aritificial Intelligence algorithms and cloud services that generate digital artifacts boosting the creation of different digital worlds.

From a semantic perspective, the term Metaverse finds its connotation within the words 'meta', aboveand 'verse', universe indicating a meta-world in which human beings—and not only—can build content and behave in ways that might not be possible in the physical world.

With a somewhat unsophisticated definition it is easy to define the Metaverse as a digital copy of the real world in which users have greater freedom and can escape the reality of life. This description could constrain the potential for future innovations as well as the development of a safe, inclusive Metaverse communities [4].

Undoubtedly trying to explore and better define the Metaverse and its opportunities and challenges can help to support the spread of knowledge around this topic to the wider community [4]. However, it is important to not underestimate the number of challenges that can evolve over the time with the Metaverse and correlated technologies development [13].

As the definition of the Metaverse and digital, virtual and immersive environments evolves, people's awareness and needs simultaneously grow and could generate a tremendous number of opportunities to solve new needs and create new technologies that will help people to experience new emotions, establish new behaviors and mature new ideas never imagined in the physical world.

Without drawing too early conclusions, the boundaries for what the ethical inferences of the Metaverse could be and how they impact people are just set by human's abilities of analyzing ethical opportunities and challenges.

What might be certain is that the Metaverse will be an entirely new world where people might spend time differently than in the physical world, where new heuristics will be developed, and new interaction paradigms will have to be designed in accordance with ethical and integrity principles.

3 Interpretations of Ethics

The term ethics, originates from the Greek word "ēthikós", meaning "relating to one's character", which itself comes from the root word êthos meaning "character, moral nature" [14].

Ethics is a part of philosophy that is concerned with behavior of individuals in society and involves systematizing, defending, and recommending concepts of right and wrong behavior and practical reasoning such as freedom, equality, duty, obligations and choice [15].

The Cambridge Dictionary of Philosophy states that the word 'ethics' is "commonly used interchangeably with 'morality' and sometimes it is used more narrowly to mean the moral principles of a particular tradition, group or individual" [16].

More broadly ethics can refer to philosophical ethics or moral philosophy and currently there are several areas of studies which can be summarized with normative ethics, applied ethics, meta-ethics.

Normative ethics studies the method to determine a moral course of action [17]. Applied ethics studies the obligations or permissions that a person has in a specific domain or context [18]. Meta-ethics studies the theoretical meaning of moral propositions and how their values can be determined [19].

A meta-ethical question is normally abstract and relates to a wide range of more specific practical questions.

A good example of a meta-ethical question is: "Is it ever possible to have a secure knowledge of what is right and wrong?".

Within the context of ethics, several correlated disciplines such as machine ethics, ethics of technology, computer ethics, and robot ethics among several others grew in the past decades.

Many of those disciplines aim to explore and investigate the ethical implications regarding different areas of technological development.

Machine ethics was coined by M. Waldrop in the 1987 AI Magazine article 'A Question of Responsibility' and focuses on the principle that intelligent machines will embody values, assumptions, and purposes, whether their programmers consciously intend to embed them into the machines or not.

As computers become more intelligent [20], it is imperative to carefully consider what built-in values might be right and what might be less purposefully right [21].

Ethics of technology is another branch of ethics that aims to address the ethical questions in relation to the technological age. This subject has been generally explored under the discipline called techno-ethics or technology of ethics [22].

Technology ethics defines the application of ethical thinking to the growing concerns that technology carry to human beings [23].

By developing theories and methods investigated in different domains, techno-ethics provides insights on ethical aspects of technological systems and practices, examines technology-related social policies and interventions, and provides guidelines for how to ethically use new advancements in technology [23].

Similarly, computer ethics aims to study the revolutionary social and ethical consequences of information technology.

The term computer ethics was first introduced by W. Maner denoting the field of philosophical inquiries dealing with ethical challenges developed by computer technologies [24].

Another field, called 'Robot ethics' refers to the ethical problems that occur with robots, such as whether robots pose a threat to humans in the long or short run.

The field of the Robot ethics or also called 'Roboethics' was built on one of the first publications setting the foundation for Robot ethics, Runaround, a short science fiction story written by Isaac Asimov [25].

All those similar disciplines of the ethics related to different technological aspects constitute a complex domain of studies and investigation dealing with the nature of responsibility, moral, safety, security and integrity.

In a similar way, with the evolution of digital, virtual, and immersive environments and the Metaverse, new ethical and integrity opportunities and challenges are rising.

Researchers, scientists, experts are in the process of jumpstarting the investigation behind what the creation of digital, virtual and imersive environments might impact to the social-behavioral sphere of human beings.

4 The Ethics of the Metaverse and Digital, Virtual and Immersive Environments: Metavethics

Businesses and innovators are currently looking into new ways of designing digital, virtual and immersive environments by considering attitudinal, behavioral and social aspects of human diversity [4].

Understanding how it is possible to create positive opportunities to deliver at scale safe, inclusive digital, virtual and immersive environments that guarantee equity and diversity in respect of ethical principles is key.

The ethics of the Metaverse and digital, virtual and immersive environments, alternatively, 'Metavethics' refers specifically to the ethics of human behavior towards the Metaverse, as it develops and grows as an advanced, pervasive immersive technology.

Metavethics concerns with the study, assessment and safeguard of the ethical, integrity and social implications of the Metaverse, digital environments, virtual spaces and augmented reality content.

While the challenges are as dated as the early attempts to create digital, virtual and immersive environments in the form of social games, such as Second life, more widespread conversations among the scientific and professional community started only in recent years.

Metavethics requires the combined expertise of specialists from numerous disciplines, who must explore, study and develop critical knowledge to inform the creation of the Metaverse and digital, virtual and immersive environments.

Metavethics arises from a need to bridge the gap between the Metaverse and the opportunities that it generates with the ethical needs and demands of people who will embark into the journey of spending time in, and creating and building new digital, virtual and immersive environments.

Some of the key tasks that specialists, scientists and experts will have to pursue are to create, amend and re-frame guidelines, standards, and policies to answer the challenges and shine the light on opportunities resulting from the scientific and technological achievements in artificial intelligence, blockchain, Web 3.0 and digital, immersive technologies such as augmented reality, virtual reality, and mixed reality.

Some of the main fields involved in the Metavethics are: computer science, artificial intelligence, philosophy, ethics, theology, biology, physiology, cognitive science, neurosciences, law, sociology, psychology, economics, industrial design and are constantly growing. Figure 1 presents the multifaceted domain of influence of Metavethics, a discipline that embraces several fields.

The knowledge generated by scientists and experts is deeply rooted in critical thinking, analysing and defining the ethical challenges that new products, features and technologies embedded in the Metaverse could generate for human beings.

These fields and further in the future, with their interconnections highlight how strongly impactful the relationship between the sociological, anthropological, and philosophical aspects that such a technology could raise in terms of ethical concerns

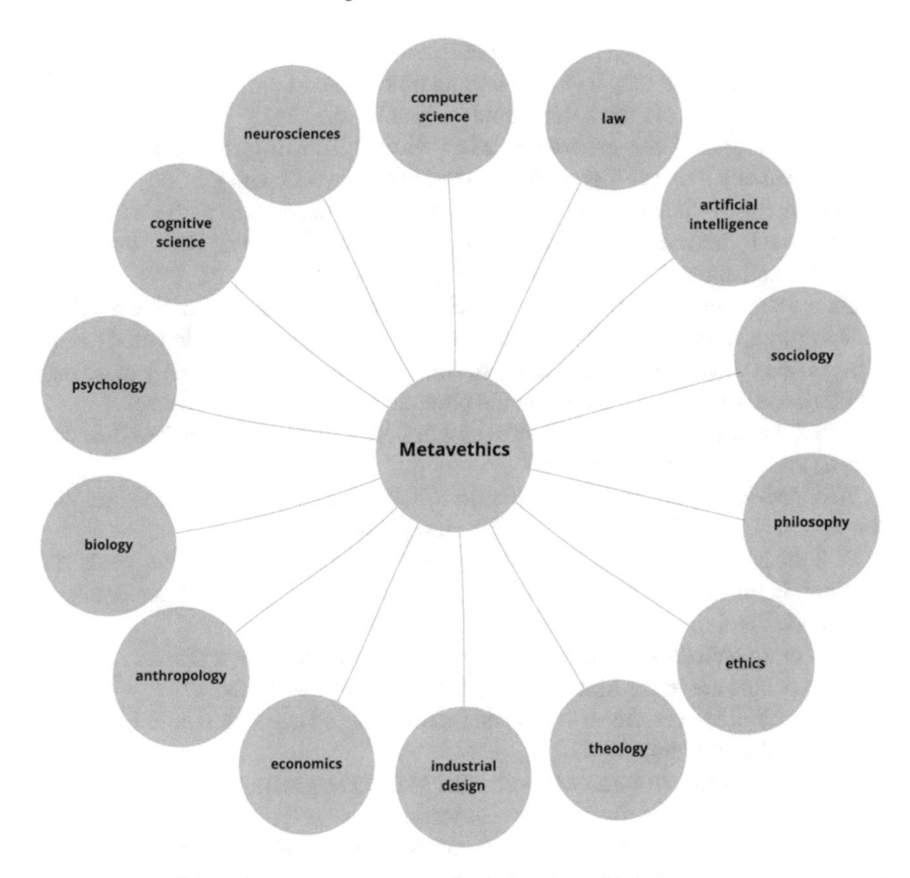

Fig. 1 The multifaceted domain of influence of Metavethics, a discipline that embraces several fields

and underline how Metavethics has the potential to serve multiple scopes, including influencing for the better the design of digital, virtual and immersive environments.

The discipline of Metavethics has the potential to help companies and organizations to identify new questions on products, technologies, services before they can be delivered to the consumers in the market and can support the community with new opportunities and ideas to preserve the physical and phycological safety as well as inclusion and privacy of every individual accessing and using the Metaverse.

Metavethics is as a multi-topic discipline that can help provide answers to a variety of different questions. One of the key questions is, to what extent the Metaverse and digital, virtual and immersive environments will provide opportunities for diminishing accessibility barriers and increasing inclusion, by guaranteeing a safe environment in which diversity and equity prevail?

Across several domains, this question can be framed with a variety of other Metavethical questions.

Regarding the sociological, anthropological level, when entering the Metaverse, how long people think they will be spending time in the Metaverse?

How often do people think they should take a break?

How intense will the experience when virtually floating in a digital, virtual environment and being still grounded in the physical world with the body?

How might the time spent in the Metaverse impact the change in the behavior?

How much the needs of accessing the Metaverse will prevail over some daily needs that constitute the behavior and personality of an individual?

On a more technical level, will the architects of the Metaverse only care about aesthetics, appearance and form given the fact that the necessity to satisfy basic human needs such as toileting, eating, sleeping, transportation and many more may not be needed (as they are framed in the physical world) in the Metaverse?

Will the architects of the Metaverse still design chairs, tables and furniture with legs, as there is no force of gravity?

Will the Metaverse become the place in which designers can practice a discipline that lays in between engineering, product design and art?

Will the Metaverse be the place where a new design approach will be created?

What heuristic will be developed to guarantee inclusion and an equitable experience for all in the Metaverse?

On a more philosophical level how is possible to define what an avatar is?

Once people are in the Metaverse, do they feel they belong there?

Who is eligible to create an avatar and for how long an avatar will be able to exist?

Will an avatar be immortal?

Will family members be able to access the Metaverse and meet a person still living there but that has passed away in the real world?

Will the content including visual art, sounds, music, verbal content, and many more, created in the Metaverse, during a meeting, a party, a social event, a gathering, be considered owned by the Metaverse, the hosting business, the avatar(s), the human being behind (if any), or a decentralized system?

Who could use that content and how it can be reshared across different Metaverses as advertising, or simply as a profitable content?

What will happen to the information about the user and the surrounding environment collected by the personal devices used to access the Metaverse?

On a more ethical level, do users of the Metaverse will be allowed to design avatars of celebrities?

What is appropriate to change after an avatar has been created by being mindful of aspect and personal privacy?

Who can access the Metaverse and create an avatar?

Will only humans be able to create their own avatars or will an AI system be able to create an AI-based avatar with human appearance?

These questions are just some of the multitude of questions that Metavethics can deal with and can dig into deeper levels to unfold the opportunities as well as challenges that arise with the design, development and release of digital, virtual and immersive environments.

Metavethics is a field of research, a discipline that has the potential to develop the needed educational effort about the Metaverse as a springboard to promote an honest conversation and disseminate and advocate the knowledge across different communities.

5 Conclusions and Future Perspectives

It is clear that allowing the community to comprehend what the Metaverse is as well as understanding the impact that safety, wellbeing, privacy, and integrity have on people are crucial aspects to provide inclusive, accessible and safe experiences for all.

Allowing people to seamlessly access the Metaverse by guaranteeing social equity and diversity and fostering opportunities to attract representations of diverse communities are the goals to design a good Metaverse and safe digital, virtual and immersive environments that answer the needs of all human beings.

With the discipline of Metavethics new opportunities to unfold the challenges, educate the community, inform people and support companies to develop safe, ethically compliant digital, virtual and immersive environments are rising.

This new field of study and domain of expertise points to the scientific community and the broader, technology-oriented community with new enquiries and inspiring chances for the creation of digital, virtual and immersive environments that are designed acknowledging positive ethical implications for human beings.

The discipline of Metavethics is still in its infancy, however it represents a corner stone upon which develop new questions, create inspiring knowledge, develop more expertise and create conversations around the complex, multifaceted concept of the ethics of human behavior towards the Metaverse and digital, virtual and immersive environments.

Metavethics is not only a discipline that has the potential to inform the design of the Metaverse but can also foster the development of new jobs and offer new opportunities to Metavethicists: the expert architects that will help organizations to build ethical, safe, inclusive, accessible digital, virtual and immersive environment and metaverses.

With this new discipline we aim to stimulate a conversation towards a future in which the ethical dimension of humans and technologies is considered and implemented while organizations are designing new digital, virtual and immersive technologies.

Metavethics proposes a common ground in which experts from different disciplines can discuss, create knowledge and disseminate it across communities, cultures, businesses.

Designing the Metaverse and digital, virtual and immersive environments is an activity that has to be done by people, for people and with people [26].

With Metavethics, it is essential to bridge the knowledge and the expertise from people who designed the physical world, together with the expertise from who is designing virtual worlds and deploy the learning and discoveries from these areas to create digital, virtual and immersive environments that are accessible, safe, secure and guarantee representation, and a culture of diversity, equity and inclusion [27, 28].

Acknowledgements This project has received funding from the European Union's Horizon 2020 research and innovation programme under the Marie Skłodowska-Curie grant agreement N∘846284.

Dr. Matteo Zallio wrote Sects. 1, 2, 3, 4, 5 and developed the infographic. Prof. John P. Clarkson co-wrote section 5 and supervised the study.

References

1. Meta: How will Metaverse change your world? Episode 5 of let me explain has answers (2022). Retrieved February 9 from https://www.facebook.com/business/news/let-me-explain-episode-metaverse
2. NVIDIA: Develop with NVIDIA Omniverse (2021). Retrieved February, 9 from https://www.nvidia.com/en-gb/omniverse/
3. Forbes: Metaverse Vs Multiverse: How To Tell Apart The 'Parallel Universe' Concepts We Can't Stop Talking About (2022). https://www.forbes.com/sites/jamiecartereurope/2022/02/07/metaverse-vs-multiverse-how-to-tell-apart-the-two-mind-bending-concepts-we-cant-stop-talking-about/
4. Zallio, M., John Clarkson, P.: Designing the metaverse: a study on inclusion, diversity, equity, accessibility and safety for digital immersive environments. Telemat. Inf., 101909 (2022). https://doi.org/10.1016/j.tele.2022.101909
5. Srivastava, L.: Mobile phones and the evolution of social behaviour. Behav. Inf. Technol. **24**(2), 111–129 (2005). https://doi.org/10.1080/01449290512331321910
6. Stephenson, N.: Snow crash. Publisher Bantam Books (1992)
7. Linden Labs (2003). Retrieved February 09 from https://www.lindenlab.com/about
8. Bobrowsky, M.: Second life founder returns to take on the metaverse. Wall Street J. (2022). https://www.wsj.com/articles/second-life-founder-returns-to-take-on-the-metaverse-11642080602
9. Warner, B.: Why Wall Street thinks the metaverse will be worth trillions. Fortune (2022). https://fortune.com/longform/wall-street-metaverse-web3-investors-roblox-meta-platforms-microsoft/
10. Facebook: Introducing Meta: A Social Technology Company (2021). https://about.fb.com/news/2021/10/facebook-company-is-now-meta/
11. Gartner: Gartner predicts 25% of people will spend at least one hour per day in the Metaverse by 2026 (2022). https://www.gartner.com/en/newsroom/press-releases/2022-02-07-gartner-predicts-25-percent-of-people-will-spend-at-least-one-hour-per-day-in-the-metaverse-by-2026
12. Dwivedi, Y.K., Hughes, L., Baabdullah, A.M., Ribeiro-Navarrete, S., Giannakis, M., Al-Debei, M.M., Dennehy, D., Metri, B., Buhalis, D., Cheung, C.M.K., Conboy, K., Doyle, R., Dubey, R., Dutot, V., Felix, R., Goyal, D.P., Gustafsson, A., Hinsch, C., Jebabli, I., Janssen, M., Kim, Y.-G., Kim, J., Koos, S., Kreps, D., Kshetri, N., Kumar, V., Ooi, K.-B., Papagiannidis, S., Pappas, I.O., Polyviou, A., Park, S.-M., Pandey, N., Queiroz, M.M., Raman, R., Rauschnabel, P.A., Shirish, A., Sigala, M., Spanaki, K., Wei-Han Tan, G., Tiwari, M.K., Viglia, G., Wamba, S.F.: Metaverse beyond the hype: multidisciplinary perspectives on emerging challenges, opportunities, and agenda for research, practice and policy. Int. J. Inf. Manag. **66**, 102542 (2022). https://doi.org/10.1016/j.ijinfomgt.2022.102542
13. Zahid, I.M., Campbell, A.G.: Metaverse as Tech for Good: Current Progress and Emerging Opportunities SSR (2022)

14. Liddell, L.H.G., Scott, R.: An Intermediate Greek-English Lexicon. Harper & Brothers (2013)
15. Snedegar, J.: Ethics and Contrastivism. Internet Encyclopedia of Philosophy (2014)
16. Deigh, J.: Ethics. In: The Cambridge Dictionary of Philosophy (1995)
17. Snedegar, J.: Contrastivism about reasons and ought. Philos. Compass **10**, 379–388 (2015). https://doi.org/10.1111/phc3.12231
18. Macklin, R.: Ethical theory and applied ethics. In: Hoffmaster, B., Freedman, B., Fraser, G. (eds.) Clinical ethics. Contemporary issues in biomedicine, ethics, and society. Humana Press (1989). https://doi.org/10.1007/978-1-4612-3708-2_6
19. Sayre-McCord, G.: Metaethics. In: Zalta, E.N., Nodelman, U. (eds.) The Stanford encyclopedia of philosophy (Spring 2023 Edition) (2023)
20. Anter, A.M., Moemen, Y.S., Darwish, A., Hassanien, A.E.: Multi-target QSAR Modelling of Chemo-genomic Data analysis based on extreme learning machine. J. Knowl. Based Syst., Elsevier, Knowl. Based Syst. **188**, 104977 (2020). https://doi.org/10.1016/j.knosys.2019.104977
21. Waldrop, M.: A question of responsibility. AI Mag **8**(1), 28–39 (1987). https://doi.org/10.1609/aimag.v8i1.572
22. Bunge, M.: Towards a technoethics. Hegeler Inst. Monist **60**(1), 96–107 (1977). https://doi.org/10.5840/monist197760134
23. Luppicini, R.: Technoethics and the Evolving Knowledge Society: Ethical Issues in Technological Design, Research, Development, and Innovation. Advances in Information Security, Privacy, and Ethics. IGI Global (2010)
24. Maner, W.: Starter Kit in Computer Ethics (1978)
25. Asimov, I.: Runaround. Doubleday (1950)
26. Auernhammer, J., Zallio, M., Domingo, L., Leifer, L.: Facets of human-centered design: the evolution of designing by, with, and for people. In: Meinel, C., Leifer, L. (Eds.), Design Thinking Research : Achieving Real Innovation, pp. 227–245. Springer International Publishing (2022). https://doi.org/10.1007/978-3-031-09297-8_12
27. Zallio, M., Clarkson, P.J.: Inclusion, diversity, equity and accessibility in the built environment: a study of architectural design practice. Build. Environ. **206**, 108352 (2021). https://doi.org/10.1016/j.buildenv.2021.108352
28. Zallio, M., Clarkson, P.J.: On Inclusion, Diversity, Equity, and Accessibility in civil engineering and architectural design. A review of assessment tools. In: Proceedings of the International Conference on Engineering Design (ICED21), Gothenburg, Sweden, 16–20 August 2021, Gothenburg (2021b)

The Implication of Metaverse in the Traditional Medical Environment and Healthcare Sector: Applications and Challenges

Mohammed A. Farahat, Ashraf Darwish, and Aboul Ella Hassanien

Abstract There are a lot of studies that have been presenting the idea of the metaverse since 2021. It's the term for the next-generation mobile computing platform, which will be extensively utilized in the future and refers to the internet accessed through VR and AR glasses. The range of illnesses people face today is different from what it was decades ago. Cancer, COPD, diabetes, heart disease, and asthma are just few of the many non-communicable diseases that pose a serious risk to human health in the modern world. As a result, efforts toward chronic disease prevention and management need to be bolstered. The emergence of metaverse theory may point healthcare in novel directions. Uneven allocation of medical resources, difficulties in follow-up, overburdening of experts, and so on remain obstacles in the management of chronic diseases in the existing healthcare system. However, advanced artificial intelligence (AI) integrated into metaverse healthcare platforms could solve these problems. The article presents the potential of Metaverse technology in the context of modern digital medicine and the future of the medical metaverse. Some related applications of Metaverse in the healthcare sector are presented. Problems and challenges are highlighted .

Mohammed A. Farahat, Ashraf Darwish, Aboul Ella Hassanien: Scientific Research Group in Egypt (SRGE)

M. A. Farahat (✉)
Computer Science Department, Higher Future Institute for Specialized Technological Studies, El Shorouk, Egypt
e-mail: Mohammed.Farahat@fa-hists.edu.eg

Information Technology and Computer Science, Nile University, Sheikh Zayed City, Egypt

A. Darwish
Faculty of Science, Helwan University, Cairo, Egypt

A. E. Hassanien
Faculty of Computers and Artificial Intelligence, Cairo University, Cairo, Egypt

A. E. Hassanien et al. (eds.), *The Future of Metaverse in the Virtual Era and Physical World*, Studies in Big Data 123, https://doi.org/10.1007/978-3-031-29132-6_7

1 Introduction

The Internet of Things in medicine, as used with augmented reality glasses, is what we call the "metaverse" in medicine. Metaverse is used in healthcare to implement the high-tech meaning of "holographic building", holographic simulation, virtual real integration, virtual-real linkage.

The concept of the metaverse must be understood in order to assess its validity and feasibility for use in medicine. The metavers2. e, the part of the internet accessible through VR and AR glasses, is becoming more and more mainstream and is seen as a forerunner of the next generation of mobile computing platforms. Similarly, the medical Metaverse might be thought of as the MIoT made possible by AR and VR goggles.

The emergence of the MIoT has provided an ideal opportunity to realize efficient and accurate graded diagnosis and treatment by connecting doctors in large hospitals (the "Cloud Experts") with doctors in smaller hospitals (the "Terminal Doctors") and to contribute to the study and improvement of related technologies.

Non-standard diagnosis and treatment, resembling that of a handicraft workshop, persists to a large extent because of the absence of real-time quality control at all times and places.

The true reason is that cloud experts and terminal doctors involved in staged diagnosis and therapy can't always easily communicate with one another because of the constraints of the Internet itself.

Therefore, it is necessary to further enhance the MIoT-based digital platforms, notably in terms of human–machine interaction and integration between the virtual and physical worlds.

There is a significant gap in China's access to healthcare resources between different regions and hospitals. Inadequate equipment coverage, insufficient technical competence, and insufficient patient satisfaction are all common problems in small rural hospitals. As a result of these "Three Deficits," many patients choose to travel to major medical centers in search of eminent specialists who can provide a more accurate diagnosis and subsequently more effective treatment, resulting in "Two Difficulties" related to hospital registration and admissions.

The influx of patients from rural areas into urban hospitals also limits the amount of time that each specialist can spend with each individual patient, resulting in what we call the "Four Limitations" in the distribution of preventative care, medical care, disease management, and rehabilitation services. As a solution to these problems, we advocated for the Internet of Things (IoT) and for the use of its three fundamental functions—in-depth perception, dependable transmission, and smart processing—to aid physicians in their daily work [1, 2].

It's encouraging that metaverse theory has been put forth as a basis for future proposals and developments in the field of medicine and healthcare; it offers a promising solution to all of these issues. Transitions toward "smart health" and "intelligent healthcare systems" are being made in the medical field. The healthcare sector is being influenced and transformed by three emerging technologies: the metaverse, artificial intelligence, and data science. It's possible to think of the metaverse as a collection of interconnected 3D MMORPGs. It is based on the combination of technologies that allow multi-sensory interaction with digital objects, digital environments, and digital people, such as virtual reality (VR) and augmented reality (AR). Even though the metaverse is just getting started, there are already many applications for 3-dimensional virtual worlds in fields like medicine, education, and training. Robotics, artificial intelligence, virtual reality, augmented reality, the Internet of medical devices, Web 3.0, the intelligent cloud, the edge, and quantum computing all come together in the metaverse of smart health to pave a new path for medical practice.

This chapter is organized as follows. Section 2 presents the basics and background. Section 3 presents and discusses the different technologies that are used with Metaverse in healthcare applications. Section 4 provides some applications of Metaverse in the medical field. Section 5 presents the problems and challenges. Section 6 concludes this chapter.

2 Basics and Background

The healthcare industry is no exception to the widespread disruption that is anticipated to result from the widespread adoption of metaverse technology. For a long time now, diagnosing medical problems, prescribing treatments, and performing surgeries have all required face-to-face meetings between patients and doctors. Telemedicine has only marginally altered this strategy. The healthcare industry stands to benefit greatly from the rapid technological advancements that have taken place over the past few years. From virtual health to mental health, and from real-world management to virtual management, the healthcare industry stands to benefit greatly from the metaverse.

Remote disease diagnosis, telemedicine, virtual health screening, and many other smart and intelligent healthcare applications are emerging thanks to the use of AI in the metaverse and data science in primary care. Among these are the provision of previously unimaginable medical and health services in a fully immersive third dimension, the remote monitoring of critically ill patients, the analysis of clinical patient data, the monitoring of blood glucose levels, the tracking of heart rates, the improvement of physical fitness tracking capabilities, and many more. Meta's acquisition of Oculus, for instance, has led to technological advancements that are aiding orthopedic surgery. In cases where breastfeeding proves difficult, Google Glass is a lifesaver. AccuVein, a startup in the medical technology industry, is using augmented reality to make the lives of nurses and patients simpler. Health care providers can also benefit from these apps and tools, which can help them do their jobs better. The use of augmented reality in surgery has the potential to greatly improve surgical efficiency. Augmented reality healthcare apps are helping doctors save lives and improve patient care in a variety of ways, from facilitating minimally invasive surgeries to pinpointing the exact location of tumors in the liver. Sync By fusing digitally enhanced images directly into the microscope of a surgical device, AR developed software to give surgeons "X-ray vision".

On the Metaverse platform, health organizations can access data from millions of user avatars. Using AI in a simulated environment to simultaneously gain insights about different treatments may be a cost-effective way to improve outcomes [3]. By using digital models, supply and demand in the healthcare industry can be better coordinated in real time. Human organs, buildings, and supply chains are all being modelled virtually. By the end of the next two decades, digital duplication will be commonplace. Information about a person's health from the time of conception until the time of their cryopreservation after clinical death will be kept indefinitely in the cloud. There will soon be a convergence of wearable devices that can hold terabytes of data per user. It's helpful to get some hands-on experience in other settings besides the OT as part of your medical education. The use of HMDs will help improve surgical precision and training. When it comes to MR, the line between the physical and digital worlds blurs. Virtual reality and head-mounted displays (HMDs) also hold great promise for telemonitoring or remote surgical guidance [4]. 3D microscopy, endoscopy, cutting-edge neuroimaging, and surgical robotics will improve surgeons' digital interactions. The learning curve is flattening as students gain a better conceptual understanding of complex anatomy and sharpen their visuospatial abilities. Immersive touch is a state-of-the-art augmented reality (AR) haptic and visualization system that utilizes human tracking and 3D visualization. By using a hologram of a virtual patient, surgeons can practice their techniques on a simulated patient [5–7]. The protection of patients' private data in the virtual world is of paramount importance. There is a pressing need for an update to the current HIPAA regulations covering the use of telehealth and mobile device integration. The patient's willingness to follow instructions and accept the treatment would also be critical. The health metaverse shows promise as a simulator for future surgical robot training. At the moment, gamification is only available in health and fitness apps. As augmented reality develops, we may see the introduction of digital trainers and new approaches to physical training. It's important to modify the ways in which data is paid for, insured, governed, protected, and secured. The user experience of teleconsultations between patients and doctors can be improved by holding them in a 3D clinic. Wearable technology like smart glasses, gloves, and sensors will collect and transmit health data in real time [8–10].

Because of these pressing problems, it's become necessary to bring medical care directly into people's homes [11]. As a result of the recent COVID-19 pandemic, the healthcare industry and its associated workforce, infrastructure, and supply chain management are under extreme strain. As a result of COVID-19, the healthcare industry has undergone rapid transformation, with stakeholders being compelled to actively pursue adaptation and innovation across the board for all relevant technologies [12–14].

In the metaverse, patients can receive healthcare services that are tailored to their specific needs while also being interactive, immersive, and recreational. Artificial

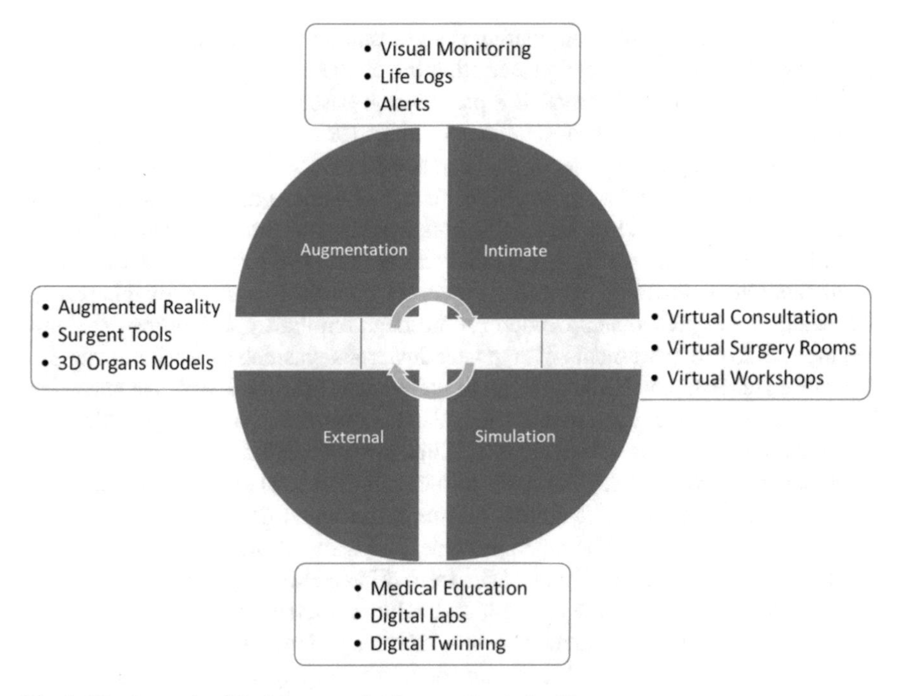

Fig. 1 The four axis of the Metaverse implementation in healthcare

intelligence (AI), augmented reality (AR), virtual reality (VR), telepresence (presence in a remote location), digital twinning, and blockchain are all examples of the Metaverse's cutting-edge technologies that are having a profound effect on medical care [15]. Better patient outcomes can be achieved at a much lower cost through the application of these technologies. Through the use of the internet, the metaverse can simulate human behavior and emotion. It encompasses the entire real-world and online economic and social structure [16]. Figure 1 shows the four dimensions of metaverse applications in medical settings.

2.1 Why Metaverse is Important in the Medical Applications

Medical practitioners can benefit from using metaverse technologies for disease prevention and diagnosis [17]. By 2020, neurosurgeons at Johns Hopkins Hospital would have successfully used an augmented reality headset made by Augmedics in a surgical procedure. Patients with chronic back pain were helped by a procedure that fused six vertebrae in their spine using a see-through eye display that projected images of their anatomy similar to an X-ray vision [18, 19]. Through the use of headsets, CT scans are converted into 3D reconstructions in the Metaverse environment, allowing for more accurate preoperative planning of surgical procedures. This aids surgeons in viewing, isolating, and manipulating specific anatomical regions during delicate procedures [20]. The enhanced prescription treatments available through the Metaverse tools for instance, EaseVR is a prescription-based solution for treating back pain using a virtual reality headset and controllers for cognitive behavioral therapy. All of the physiological aspects of pain can be addressed with the aid of these techniques, which include inducing deep relaxation, redirecting attention, and increasing interoceptive awareness. Reconstructing human body parts through plastic surgery is a highly intricate process [21, 22]. Virtual reality (VR) in the Metaverse may play a significant role in the future of plastic surgery if the virtual avatar can reliably predict the results of a hypothetical procedure. A thorough familiarity with human anatomy and the skill to use instruments with greater dexterous grasping capability, as well as flexibility and individual adjustment, are prerequisites for the Metaverse surgeries. Surgery of any complexity, from the removal of tumors to delicate spine procedures, is within reach [23]. Advanced capabilities in image visualization could be unlocked by the metaverse in the radiology domain, allowing radiologists to see dynamic images in greater detail for better diagnosis and more informed decision-making. Better radiology education could be made possible, and teams could pool their resources to analyze the same 3D medical images from different locations, saving both time and money [24]. Through the use of high-quality immersive content and gamification features, the healthcare metaverse may increase patient engagement by assisting clinicians in clarifying difficult concepts, demonstrating upcoming procedures, and checking that patients are taking their medications as directed. Patients' vitals, CT scans, health records, and genetic test results are all combined to create a digital simulation of the patient's anatomy and physiology, which is then used to monitor and gain insights into the patient's health condition. This is how digital twin solutions in the Metaverse will keep patients informed and involved in their treatment. Using the virtual dashboard, patients can see their health information, which can facilitate discussions with their doctors, researchers, dietitians, and other interested parties to better tailor their care. The recent pandemic has increased demand for telemedicine services, and the Metaverse may be able to offer a more immersive experience than traditional video conferencing-based telemedicine systems [24]. AR glasses could be used to access live videos and audio communications within the Metaverse systems, allowing patients and doctors to have real-time conversations. Respondents would be able to interact with each other in real time, and remote clinicians would be able to watch live feeds of critical situations in order to respond quickly and effectively [18].

The metaverse could significantly alter the way doctors are educated and trained. Rather than spreading theoretical knowledge, AR creates an environment in which actual techniques can be explained. Reputable institutions are gradually adopting the use of virtual reality (VR), augmented reality (AR), mixed reality (MR), and artificial intelligence (AI) to educate doctors by simulating complex real-time procedures and providing information on the cellular level details of the human anatomy [25]. Possible medical uses for the metaverse could be broken down into four distinct groups, each of which is dependent on a different set of enabling technologies: augmented reality (AR), lifelogging, virtual reality (VR), and the mirror world. An augmented reality T-shirt, for instance, may let students clearly visualize the human body in an anatomy lab. If a mirror world were implemented, it would accurately mimic the real world by including all relevant data regarding the surroundings, as gathered through the use of digital maps and simulations. The present pandemic emergency has greatly boosted the utilization of the Metaverse. People from all around the world can interact in the mirror world through a standard mirror world platform and play games together. A similar concept in the medical field has players contributing to studies in the scientific community. The metaverse can be the most effective surgical training tool by facilitating the highest level of cooperation and immersion thanks to its ability to deliver 360-degree visualizations of the body's illnesses. While still in the testing phase, this technology has the potential to revolutionize medical education and training.

3 Transforming Healthcare with Metaverse Technologies

Extended reality, blockchain, AI, IoT, 5G and beyond, digital twin, big data, quantum computing, human–computer interaction, computer vision, edge computing, and 3D modelling are all discussed in length in this section as they pertain to the enabling technologies of the Metaverse for healthcare. Figure 2 shows an example of the aforementioned Metaverse health-care enabling technology.

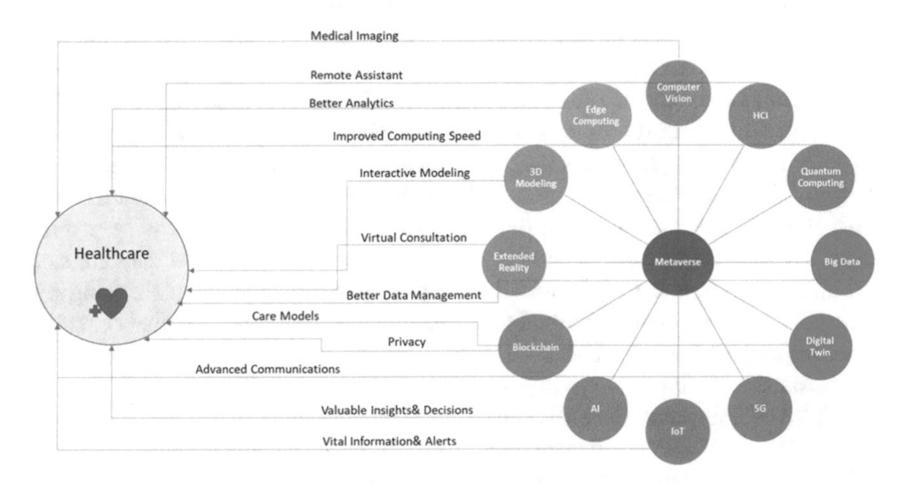

Fig. 2 The healthcare with different Metaverse technologies

3.1 *Extended Reality*

With the use of AI, computer vision, and linked devices like smartphones, wearables, and HUDs, extended reality may be achieved. By merging speech, gestures, motion tracking, vision, and haptics, this new technology is revolutionizing the way services are given, boosting the quality in numerous areas. Conventional wisdom held that XR would only have use in the entertainment sector. It was assumed that giving users a more enveloping experience would only improve their enjoyment of media like video games and movies. However, XR usage has considerably exceeded these forecasts. An ever-expanding range of sectors, from healthcare to industry, are adopting it [26]. Metaverse is where XR will mature into its full potential. VR and AR technology is integrated to create a sense of virtual presence in the Metaverse. With the help of the Metaverse, users may have realistic simulations of the activities they like in the real world, but in a digital setting. Increased interest in XR in healthcare throughout the world is a direct result of the wide spread of COVID-19 pandemic [27]. This increase in popularity can be partially attributed to the growing interest in remote medical diagnosis and treatment. Reimagining medicine using XR in the Metaverse. Medical students can learn more in real-life settings than in a typical classroom. However, in the real world, even a minor slip-up might have catastrophic results for one person. In this case, students of medicine will be able to hone their skills in a virtual 3D environment that is just as lifelike as the actual thing, thanks to XR applications in the Metaverse [28]. Surgeons will benefit from this Metaverse-based interactive virtual 3D environment as well [29]. Surgeons may now see organs, tumors, X-rays, and ultrasounds in real-time and from many angles thanks to the Metaverse and XR, all without taking their eyes off the patient. The effectiveness and speed of therapy are both improved with the use of 3D patient models available in the Metaverse [30]. Doctors can instruct laypeople on how to do CPR in a simulated setting utilizing XR technology in the Metaverse [31] when real-world events make it necessary for them to do so. Patient rehabilitation in the areas of physical health, communication skills training, and mental health may all be accomplished through the Metaverse using XR and remote-controlled equipment. Since patients' medical data are safe and easily accessible in the Metaverse, doctors may review them remotely using XR devices and provide more informed drug recommendations during virtual consultations. Although XR-enabled Metaverses help provide better healthcare in a range of places, substantial hurdles exist in putting them into practice. There is a risk of a data breach in the Metaverse since XR-related technologies may access a lot of personal information about patients; this might have serious consequences for patients' health and privacy, as well as for the doctor's professional standing. It may be quite pricey to create and install XR technologies that help with healthcare delivery in the Metaverse. Since not all patients can afford or benefit from this cutting-edge equipment, it complicates medical visits.

3.2 Blockchain

In a white paper published in 2008, Satoshi Nakamoto [32] laid forth the principles around which blockchain technology is based. A blockchain is a distributed digital ledger of all the transactions that have ever taken place on the blockchain network. All participants' ledgers are updated whenever a new block is added to the chain [33, 34]; each block comprises numerous transactions. Distributed ledger technology is another name for blockchain since it is managed by more than one entity (DLT). A blockchain is a distributed ledger technology (DLT) that uses a cryptographic hash signature to permanently record transactions. Since this is the case, it follows that it would be readily obvious if even one block in a chain were to be tampered with. To compromise a blockchain, hackers would need to alter each block in the chain on all of the distributed copies of the ledger. Decentralized digital assets may be issued, owned, and used with the help of non-fungible tokens and cryptocurrencies on the blockchain [35]. Since centralized data storage has several issues with security, privacy, and data transparency, the notion of the Metaverse would be lacking without blockchain. The blockchain will transform the Metaverse into a decentralized digital source, making it cross-platform and universally accessible. Without needing to go via any kind of governing body, users of the blockchain-based Metaverse can enter whatever kind of digital world they choose. Through the use of blockchain technology, the Metaverse [36] will be able to more easily collect data that is both authentic and of high quality. The blockchain's immutability and openness will guarantee the safekeeping of the Metaverse's massive data [37]. It is expected that the availability of medical professionals will increase as a result of the Metaverse's implementation. Patients will have more freedom and a more comfortable setting in which to talk to their doctors in the Metaverse [38]. This minimizes the time necessary to acquire a diagnosis and consultation. The blockchain-enabled Metaverse will give exact patient Enhanced Medical Simulations data to physicians, allowing them to make more accurate judgements. The blockchain will prevent such sensitive information from being updated or tampered with by attackers [39]. There are also certain limits to the blockchain-enabled the Metaverse for healthcare. The high price tag and extensive infrastructure needs of this technology pose obstacles to its widespread implementation. Patients are at danger since this technology is unrestrained and unregulated. The technology's intricacy will make it difficult for end users or patients to adapt to it. Since all data must be stored at each node in the chain, the blockchain-enabled Metaverse might be painfully sluggish [37]. Small hospital chains cannot adopt this technology because of its enormous energy use and complexity.

3.3 Artificial Intelligence

AI, often known as machine intelligence, is the study, creation, and management of machines that can learn to act intelligently on behalf of people in a wide range of situations [40, 41]. Machine learning, computer vision, NLU, and NLP are all subsets of AI, as are any other applications or hardware that aid in these areas of study. Artificial intelligence will be used to improve the Metaverse's foundation, leading to a more immersive 3D experience and better in-game functions. The quality of services and the Metaverse ecosystem as a whole will both rise thanks to advancements in AI technology. To improve medical device efficacy, lower healthcare costs, improve healthcare operations, and increase access to medical care, the health industry has recently begun using revolutionary techniques like augmented reality (XR) and big data combined with artificial intelligence (AI) in software and hardware [42]. The metaverse facilitates the in-depth study of medical concerns and the exchange of relevant data between doctors and their patients. Artificial intelligence is being used to examine medical records and make diagnoses. The Metaverse, aided by AI, provides clinicians with the high-quality 3D photographs and scans of patients necessary for intervention. Medical professionals may benefit greatly from the insights provided by AI, which can aid in patient prioritization, reduce the likelihood of making mistakes when examining electronic health information, and improve the quality of their diagnosis [43]. Large amounts of health information and patient records make it challenging for doctors to keep up with the newest medical breakthroughs and offer high-quality care that is focused on their patients. Artificial intelligence algorithms in the Metaverse can instantly scan electronic health records and biological data gathered by medical units and experts to provide instant and reliable suggestions to clinicians [44]. Drug development, disease prognosis, and disaster relief are all domains where the metaverse and AI might work together to benefit humanity. While it can provide new insights and speed up interactions with healthcare data for both patients and physicians, an AI-enabled metaverse also poses substantial threats to patient privacy and ethical problems and can even produce medical blunders that can mislead doctors in giving therapy. Adoption of AI-enabled the Metaverse will be hampered by the inability to provide enough rationale for findings [45, 46].

3.4 Internet of Things

Many billions of devices are already online and sharing data under the umbrella term "Internet of Things" (IoT) [47]. The widespread availability of wireless networks and the advent of cheap computer processors have made it possible to link together hitherto disparate systems, from handheld devices to interplanetary operations. These gadgets include sensors and the ability to connect with one another, so they can exchange data in real time without a human overseer [48]. With the aid of the Metaverse, the Internet of Things (IoT) will merge the digital and physical worlds into one that is both more intelligent and sensitive to human needs. IoT-enabled Metaverse applications have found broad use in healthcare, particularly remote patient monitoring. A patient's vital health data, such as heart rate, blood pressure, temperature, and more, may be automatically gathered by IoT devices and shown in the Metaverse's 3D environment, even if the patient is not physically present at the hospital or clinic. As a result, patients no longer have to go out of their way to see doctors or get their medical records [49]. Surgeons will be able to undertake intricate surgeries that would be difficult for human hands alone by deploying small Internet-connected nanobots within the human body and monitoring them in a virtual 3D environment provided by the Metaverse. Simultaneously, tiny IoT devices or nanobots used in robotic surgery might lessen the size of incisions. Due to the reduced level of trauma, patients can recover more rapidly. Management of chronic conditions, including better sleep, prescription reorders, and notifications in the event of medical crises, can benefit from the combined efforts of the Metaverse and the Internet of Things as well. Healthcare is a sector in which data security and privacy concerns are of paramount importance [50–52]. As IoT-enabled devices don't conform with data protocols and standards in real time, data security and privacy concerns will be a worry of the IoT-enabled Metaverse. It will also be difficult to combine IoMT gadgets with Metaverse. There will be challenges in aggregating data for critical insights and analysis in the IoT-enabled Metaverse due to the lack of standardization in data and communication protocols. Another obstacle to implementing Internet of Things (IoT)-enabled the Metaverse in healthcare is the high price tag associated with the necessary hardware and software.

3.5 5G & Beyond

"5G and beyond" refers to wireless technologies beyond the fifth generation. In comparison to 4G LTE networks, it is faster, has lower latency, and can handle more users. 5G technology can achieve speeds of more than 20 GB/s, much beyond the highest speed of 4G (1 GB/s). And it reduces latency, which might boost the efficiency of business apps and other digital experiences [53, 54]. With its faster speeds and improved connectivity, companies will be able to get more done. From improving smart healthcare to building smart cities, all fields will benefit from networks that extend beyond 5G. The full potential of the metaverse will be realized with the help of 5G and other future technologies. Low latency and high bandwidth connectivity, as given by 5G and beyond technologies, are essential for the development of rich media content employing immersive technologies like augmented reality (AR), virtual reality (VR), and the internet of things (IoT) in the Metaverse. So, the ease with which new types of content may be created in the metaverse will be greatly influenced by 5G and beyond technologies [55]. The healthcare industry stands to gain from the innovations made possible by the metaverse. By leveraging 5G and beyond technology, the Metaverse can deliver a high-quality immersive experience that will benefit both patients and doctors. Many facets of healthcare, such as virtual wellness, mental health, and sophisticated surgical procedures, will benefit from the high speed and low latency connectivity made possible by 5G and beyond technologies to the Meta-verse [56]. Metaverse advancements made possible by 5G and beyond technology will allow for uninterrupted, immersive medical education and training for future doctors. With the laggy connections typical of today's wireless networks, doing remote surgery in the 3D world afforded by the Metaverse is a risky proposition at best. If this takes too long, harm might come to the patients. By eliminating the time gap between a remote surgeon's input and real-time action, 5G and beyond networks promise to make remote operations safer [57]. In the medical profession, for example, 5G and beyond enable the Metaverse to deliver quicker and latency-free connectivity, but the cost will rise as a result of the adaptation of these technologies. In addition, there will be a greater volume of data flowing in, necessitating substantial increases in both processing power and data storage capacity. Metaverse use in healthcare and other sectors will be hampered by the lack of 5G-enabled equipment.

3.6 Digital Twins

A "digital twin" is the virtual representation of a physical item or process that can replicate its physical counterpart in real time [58]. The term "digital twin" was first used in 1991 in David Gelernter's book Mirror Worlds [59]. Even Michael Grieves, writing in 2002 [60], alludes to the idea. It was NASA who came up with the first practical definition of a digital twin in 2010 with the aim of improving physical-model simulation of spacecraft [61]. The development of digital twins can be attributed to the maturation of product development and engineering methods. A "digital twin" is an electronic copy of a physical product, service, or operation. A digital twin is a virtual replica of a real-world object, which can be anything from a piece of machinery to a medical gadget or even something as huge as a city. The Metaverse, on the other hand, is a technology that creates a virtual world where everything and everyone behaves just like they do in the real one. In the metaverse, everything has a digital twin that serves as a copy of the original [62]. In order to evaluate the needs for operational strategies, personnel, and care models, a digital twin of the whole hospital can be created in the metaverse. These virtual models in the metaverse can be useful in addressing issues like bed shortages, the spread of disease, the scheduling of medical staff, and the accessibility of operating rooms. The use of digital twins provided by Metaverse can enhance patient care, save expenses, and boost staff efficiency. This is essential for making strategic decisions in the healthcare industry, which is both complex and delicate. The use of digital twins allows for a completely virtual, risk-free hospital to be created in the metaverse. The Metaverse, made possible by digital twins, will also aid in the development of patient-specific artificial organs [63]. Brain and heart surgeons can use the Metaverse's digital twin capabilities to practice complicated procedures virtually before attempting them in real life. Developing a digital duplicate for every physical item in the world is a formidable task that calls for massive amounts of data and processing power. The creation of a virtual organ or living being is the greatest difficulty in the metaverse. A significant difficulty for healthcare applications of the digital twin-enabled Metaverse is recreating a human being with all their organs and systems functioning in perfect harmony in real time.

3.7 Big Data

Information asset characterized by such high volume, velocity, and variety as to necessitate specialized technology and analytical methods for its translation into value" is how the term "big data" is formally defined [64]. Such massive amounts of data acquired from several sources at rapid speeds may be put to good use through analysis. With improved sentiment and behavioral analytics, predictive assistance, and fraud detection, big data is finding applications in a wide range of sectors, including healthcare, manufacturing, education, agriculture, and more. However, the metaverse makes it possible for individuals to engage with and explore a virtual world in a way that is permanent, real-time, and interoperable. The metaverse has replaced big data as the leading edge of computing. Because of the metaverse's interference, big data is fast expanding and gaining traction as a crucial technology. Traditional data analytics tools have trouble dealing with the variety of data types present in the metaverse, which encompasses organized, semi-structured, and unstructured forms. Integrating big data analytics with the metaverse, on the other hand, will help businesses create a unified point cloud platform for collecting and analyzing both internal and external data in order to draw useful conclusions. Given the necessity to make instantaneous judgements and the need to be able to reliably anticipate the future, this quality is crucial in the metaverse. Although the metaverse and its underlying systems create massive amounts of data, which may be put to good use through a variety of analytic methods, it is imperative to have effective big data tools at hand to deal with this deluge of information. As computerized patient simulations become more commonplace in healthcare, this becomes increasingly crucial for revealing previously hidden patterns [65]. If the metaverse is used in medicine, it can accurately predict the outcome of a therapy and the chance of its success. Knowing how the target audience will react to present or future decisions is made much easier with the integration of big data analytics technologies in such situations. Many scholars are interested in merging big data technology with the metaverse, and a few publications on the topic have been published. The authors of [66] provide such research, which describes the Edu-Metaverse and its many educational applications in depth. Similar to this, the authors of [67] have detailed the obstacles and possibilities inherent to utilizing the metaverse and big data in Indonesian corporate operations. Although study of the big data enabled-Metaverse is in its infancy, it is becoming increasingly important as the Metaverse has the potential to generate vast amounts of real-time data in a wide variety of formats, all of which can be efficiently managed by combining a number of big data technologies.

3.8 Quantum Computing

To do calculations, quantum computers make use of quantum mechanical phenomena including superposition, interference, and entanglement. Quantum computers [68] are machines capable of doing calculations using quantum mechanics. The computational tasks that can be taken on by quantum computers are vast. Critical choices may be made with the aid of quantum computing far more quickly than with traditional computing. In light of this, quantum computing is becoming increasingly popular in a wide range of industries, such as finance and medicine [69, 70]. The goal of the metaverse, on the other hand, is to create a virtual environment that closely reflects its real-world analogue by combining many different sectors into a single ecosystem. The objective is to produce a user experience that combines elements of social media, gaming, and simulation into a virtual world that is a close approximation of the actual thing. Particularly, the pandemic's effect on medical advice has led to the rise of digital healthcare. The Metaverse's high-quality images and attention to realism will be put to use in creating lifelike interactions between patients and doctors. Using this function of the Metaverse, patients and doctors will be able to interact often in this virtual space. These virtual worlds need massive amounts of computing power to create. The metaverse contributes to the development of a realistic, interactive healthcare virtual community. Overcoming the massive obstacles of security, high processing capacity, and cyber-attacks will be feasible with the help of the metaverse provided by quantum computing. Quantum computing will offer the necessary security for the Metaverse through the creation of quantum-enabled security applications. As the scope of the metaverse widens, the use of quantum-resistant security measures will become mandatory for all operations and transactions in the medical profession. Medical applications will greatly benefit from the vast processing power made available by the Metaverse's implementation of quantum computing. Metaverse healthcare apps heavily rely on computational and simulated data. As a result, both the efficiency of computation and the quality of care provided by medical professionals are enhanced by the advent of quantum computing. In order to guarantee that the rules in place prohibit residents and algorithms from manipulating the healthcare apps in the Metaverse, developers can use quantum randomness. Although quantum computers have shown promise, they have yet to become the industry standard. Existing technologies and techniques for enabling the metaverse are based on binary systems, whereas qubits are used in quantum computing. This creates a substantial barrier to the smooth operation of quantum computers in the metaverse. The high energy requirements of quantum computing are an issue that has to be fixed before its full potential can be realized in the context of the metaverse. All of the Metaverse's security measures would be rendered useless if quantum supremacy fell into the wrong hands.

3.9 Human–Computer Interaction

Studies in human–computer interaction (HCI) draw from a wide range of academic disciplines because they address such broad questions as how best to build computer technology to facilitate communication between humans and machines. Textual or display-based control has been supplanted as the main paradigm in HCI [71] by voice, gesture, visual, and brain signal interaction. The development of the metaverse will be made possible by HCI, VR, AR, and emerging content production and collaboration technologies. Wearable consumer head-mounted displays are the HCI technology that will power the visual interactions in the metaverse (HMD). In the metaverse, these HMDs will be important in facilitating interaction between people and their environments. The senses of users in the metaverse will be activated by these HCI gadgets. Users will be able to employ a haptic wearable device, another kind of HCI technology, to get a sense of touch, smell, and taste while in the metaverse [72]. Using the Metaverse's realistic environment, these gadgets will also allow humans to collaborate with robots in far-flung locales. The HCI-enabled metaverse will have technologies including holographic construction, holographic simulation, augmented reality integration, and augmented reality interconnection. There are several ways in which healthcare might benefit from the HCI-enabled metaverse. With the enhanced interaction and immersion given by the HCI-enabled Metaverse, medical education for students will be facilitated, research findings will be more widely disseminated, a consultation will be enhanced, and accurate, graded diagnosis and treatment will be made available. HCI-enabled Metaverse users will have access to enhanced medical care, including but not limited to: disease prevention; telehealthcare; physical examination; disease diagnosis and treatment; rehabilitation; chronic disease management; in-home care; and first aid. Surgeons may execute the surgeries from afar with the aid of robots in the Metaverse's realistic setting. Human–computer interaction (HCI) technologies are useful in many ways for the metaverse, but they also present many difficulties. In order for the Metaverse to reach its full potential, it will take a long time for modern HCI technologies like head-mounted displays and haptic wearable devices to mature. Incompatibility with established standards and conventions makes it difficult to incorporate such HCI gadgets into the metaverse. It will be more challenging for consumers to adjust to these devices since they are too expensive for the average consumer.

3.9.1 Computer Vision

The field of computer vision analyses how computers see and make sense of moving pictures and still images. Perceiving a visual signal, understanding what is being seen, and extracting complex information in a form usable by other processes—all of these are tasks performed by biological vision systems and fall under the umbrella of computer vision. The field of computer vision is concerned with simulating and automating key components of the human visual system via the use of sensors and machine learning algorithms [73]. Computer vision is used in many different areas, but some of the more well-known ones include autonomous cars, facial identification, picture search, and object recognition. Using technologies including object recognition, plane detection, facial recognition, and movement tracking, computer vision contributes to the Metaverse's replication of the user's physical experience in the 3D virtual environment. The user's location and movement inside the metaverse may be ascertained with the use of computer vision. To represent its users, the Metaverse will employ computer vision to create avatars. The user's position and motion in the metaverse will be monitored in real time by an interactive computer vision system. In the metaverse, interaction with the environment is mandatory. Therefore, the use of computer vision for picture processing is crucial to the creation of more compelling Metaverse experiences. Using computer vision, the metaverse may be linked to the actual world in real time. The user will become irritated if the produced avatar does not work with them in all circumstances. By using computer vision, the Metaverse can provide users with a realistic representation of the 3D virtual world. By connecting IoMT devices to the Metaverse's computer vision system, medical imaging can be improved. It may be used to monitor the growth of a tumor and observe cancer cells in three dimensions. As a result, it can be useful for the education of future doctors by providing a wide range of practice situations. By facilitating in-house sickness detection and decreasing worldwide dissemination, the computer vision-enabled Metaverse can help prevent pandemics like COVID-19. As a computer vision-enabled metaverse, it also aids in remote patient monitoring and makes choices based on critical data. There are certain obstacles to overcome when combining computer vision and the Metaverse. Unfortunately, the computer vision-enabled Metaverse's projection methods aren't always spot-on, which can lead to distorted environments. When utilizing a distorted image display, visual symptoms may occur for users who have presbyopia, nearsightedness, farsightedness, or other age-related or hereditary eye conditions that affect their ability to concentrate or coordinate their eyes. Surgery and treatment for some disorders can benefit greatly from the Metaverse's computer vision capabilities, yet even the tiniest error might have catastrophic results.

3.9.2 Edge Computing

Edge computing is a style of distributed computing that does computation and stores data in proximity to the data's original sources. By using computational resources closer to the user, reaction times may be slashed and bandwidth used more efficiently [74]. Bandwidth and latency concerns arise when all device data must be sent to a single location, such as a data center or the cloud. Since edge computing processes and analyses data at its source, it is more efficient. When data doesn't have to be sent to a remote server or data center, network latency is drastically reduced [75]. The faster and more complete data processing made possible by edge computing, especially mobile edge computing on 5G and beyond networks, enables more in-depth insights, faster reaction times, and better user experiences. The Metaverse is a future virtual environment that will combine real-world elements. Gathering this kind of information requires a massive amount of data from several sources. Ultimately, the produced information must be uploaded to some kind of central repository or cloud service. There will be a massive increase in the number of connected devices and the volume of data sent between them. The Metaverse apps will crash and burn if there is even a momentary hiccup in the transmission or storage of data. Edge computing will increase the Metaverse's transmission speed while reducing bandwidth consumption [76]. By doing so, the creation of the metaverse is guaranteed. With edge computing, the Metaverse can increase throughput and keep only a small amount of data on hand. Edge computing in the Metaverse places computer resources closer to where data is generated. Using this method, health apps in the metaverse may make better use of edge computing to improve data collection, storage, and analysis. Wearables like trackers and smart watches will be able to effectively transfer data and offer doctors an up-to-date status of critical patient vitals like heart rate and blood pressure, therefore alerting medical personnel to possible issues before they develop. By collecting patient data and triggering actions in real-time based on the results, health monitoring systems can enable remote patient care. With the help of imaging models, potential problems in X-rays may be identified in real-time, allowing doctors and radiologists to focus on the most pressing cases while remaining in their remote locations. The Metaverse's ability to retrieve data in real-time will be improved through edge computing. It also requires the difficult installation of many new edge devices on the current network. In addition, it raises a number of issues about safety, data loss, device diversity, and financial burdens [77].

3.9.3 3D Modeling

The term "3D modeling" refers to the act of creating a representation of any three-dimensional surface of an object using specialized 3D modeling techniques based on mathematical coordinates. Improvements in image processing, CAD, and modeling techniques have made it possible to construct 3D models that are both highly realistic and reliable [78]. 3D modeling is used in many fields, from film and television to video games and even interior design and architecture. The success of the Metaverse will depend on how accurately 3D models of real-world items can be recreated in the virtual environment. Everything from individuals to towns to rainforests is available in 3D in the Metaverse. When it comes to the Metaverse, 3D modelling will be the backbone of all the apps. In the medical field, 3D modeling-enabled Metaverse might be used to construct interactive anatomical representations from high-quality images. Viewing medical aid goods in three dimensions helps patients make informed decisions about which tools they need. Medical professionals can benefit from the metaverse's 3D modelling capabilities by better understanding a patient's disease. These results will help medical professionals better care for their patients and provide their therapies with more accuracy. Researchers in the medical field may use the 3D modelling capabilities of the metaverse to design artificial organs and test how well they function in different virtual settings. The entire potential of the metaverse can only be realized with the help of 3D modelling, yet this shift is fraught with challenges. It takes a lot of time and effort, both human and mechanical, to complete this journey. The Metaverse is an ever-expanding digital universe with no limits on content or development, posing a continuing challenge to 3D modelling methods. Dynamic 3D models are required because the Metaverse requires dynamic objects.

4 The Use of Metaverse in Medical Applications

When compared to the "handicraft workshop model" of healthcare, in which diagnosis and treatment vary from doctor to doctor and hospital to hospital, the metaverse in medicine greatly aids in delivering complete healthcare. Decisions in a holistic healthcare scenario will be informed by both the expert's recommendations and the data gleaned from the different Metaverse enablers. There are many uses for the metaverse in healthcare, from research and diagnosis to physical examinations and even insurance. Virtual physiotherapy, biopsy, counselling, and alarm response are all examples of Metaverse applications that might see significant growth in the near future. In order to describe tissues, a non-invasive method called virtual biopsy can be used to capture and analyze images of the target area. Patients in need of physiotherapy during their recovery process might benefit from virtual physiotherapy by being led through exercises and movements by a computer. Applications of the Metaverse in all-encompassing healthcare are discussed here, as seen in Fig. 3.

4.1 Medical Diagnosis

Diagnosing a patient's condition is the first step in treating their illness. With the support of cutting-edge innovations like augmented and virtual reality (AR/VR) enabled MIoT models, an extended digital twin, blockchain, 5G, and many more, the adoption of the Metaverse in healthcare greatly aids in the efficient diagnosis of a patient's medical issues. Additionally, the Metaverse may be seen as an improvement over the current medical IoT by removing barriers to human–computer contact, interconnection, and integration between the actual and virtual worlds. Holographic creation, emulation, real and virtual world interaction, and integration are all made easier with the use of the Internet of Things (IoT), which can be accessed through AR or VR glasses, therefore easing the complicated difficulties faced in a healthcare setting.

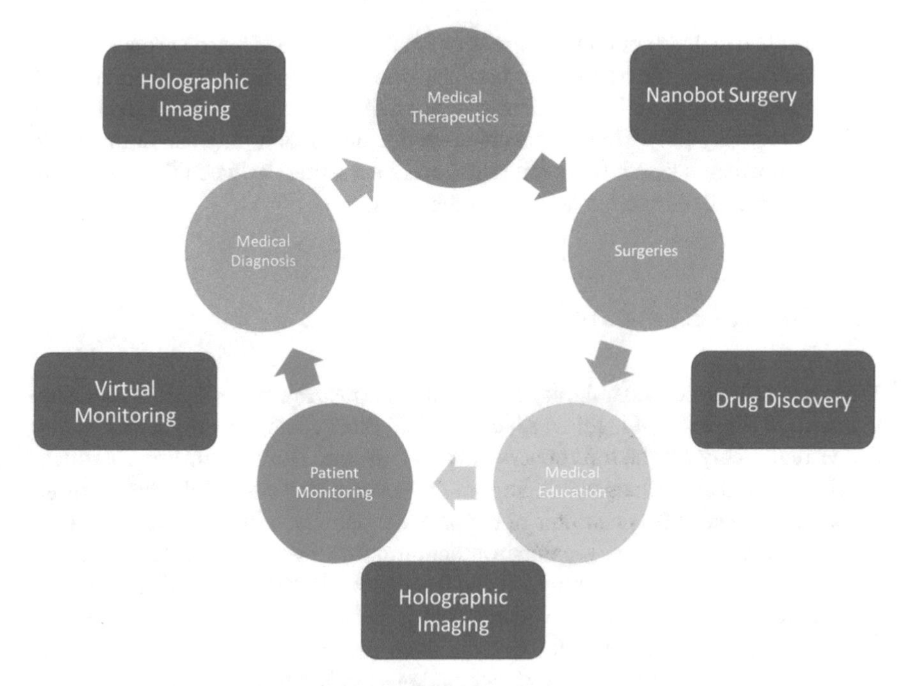

Fig. 3 The healthcare implications in the Metaverse

4.2 *Patient's Monitoring*

Medically, the Metaverse will be of great use as telepresence, digital twins, and blockchain all come together to improve patient monitoring. Remote medical care is provided using telepresence, often known as telemedicine [79]. A "digital twin" is a replica of a physical thing or person that may be studied virtually to get insight into the original. In emergency scenarios, these patient mannequin test subjects can be used to predict how a patient would react to a surgery or medication long before those procedures or medications are ever administered to real people. Given the critical nature of medical records, blockchain technologies provide a safe and reliable means of storing and transmitting these records, reducing the likelihood of any tampering or other compromise.

If these three elements are properly coordinated, patient monitoring may be provided efficiently. The Metaverse saves the day by fusing all of these technologies into one coherent whole. The implementation of CoViD-19 has prompted health-care providers to consider how they might provide high-quality treatment to patients without physically being present in the same room [80, 81]. With the advent of the Metaverse, however, the medical field will be able to treat patients across continents without ever having to leave the comfort of their own homes. Virtual reality and augmented reality processes provide a feeling of "being there," which is useful in the case of patient monitoring. These implications extend beyond only the doctor-patient relationship to include communication between the patient and his loved ones as well. With the support of the Metaverse, families who are geographically separated can feel as though they are all in the same room, which can have a positive effect on the patient's recovery [82]. Thus, metaverse-based patient monitoring can dramatically enhance a patient's health by facilitating good interactions between the patient, his healthcare providers, and his family members.

4.3 Medical Education

When it comes to medical training, the advent of the metaverse represents a watershed moment. In the realm of medical education, IoT, blockchain, AI, augmented reality, and virtual reality are the forerunners of the metaverse. Through the use of artificial intelligence and blockchain technology, the Metaverse is able to build a digital virtual world that exceeds the constraints of the physical one. Even in a high-stress clinical setting, medical students may maintain attention and engagement with the session, the other students, and the instructor by using these tools. Traditional medical education involves a teacher taking a class of students to see a patient, at which point the teacher presents the patient's case and the class discusses the relevant medical literature. The advent of digital integration and 3D technology has led to a revolutionary shift in clinical education, with the patient now being transported virtually to a class of medical students. Virtual reality is being used for a wide variety of medical purposes outside of just surgery, such as colonoscopy examinations, student instruction, and more. Improving patient care requires an innate familiarity with anatomy and the ability to engage with it. Virtual reality (VR) is used to teach students and many practicing physicians about the human body's anatomy. It is possible to study the human body in its entirety, from the bones and nervous system to the muscles and everything in between. This type of education opens up previously unattainable doors and produces superior doctors for the future. Surgeons doing augmented reality procedures utilize a headgear with a transparent eye display to examine pictures of the patient's interior anatomy, such as bones and other tissue, generated from CT scans. Virtual reality (VR) allows the user to generate a patient-specific 360-degree anatomy reconstruction and present the surgical plan to patients. With this method, you may better educate patients about their condition and win their confidence.

The COVID-19 epidemic has affected the instruction of the medical students and practitioners considerably. The quantity of available surgeons, the volume of patients, and the availability of in-person training opportunities were all impacted by the pandemic scenario for various reasons [83]. The term "Mirror Reality" (MW) refers to an artificial simulation of our physical world. MW is an independent virtual environment built on the blockchain, and it represents the information and appearance associated with an application. As a result of the MW, everyday life is simplified and productivity is increased. Digital labs and virtual educational environments are two types of mirror worlds that are particularly well-suited for educational purposes. Example uses of MW in the realm of medical education include "virtual educational environments" [84].

4.4 *Surgeries*

The metaverse is rapidly becoming a vital tool for surgeons and other medical professionals. Virtual reality (VR) headsets and haptic gloves are only two examples of the instruments presently being used by surgeons to simulate actual surgical operations, improving both readiness and efficiency. Surgeons can benefit from augmented reality since it allows them easy access to data during procedures. Surgeons may have quick, simple, and hands-free access to patient data using augmented reality by superimposing 3D virtual models over the patient's body. Professors and teachers might use the metaverse to demonstrate complex operations in three dimensions in the Metaverse.

4.5 *Medical Therapeutics and Theragnostic*

Therapeutics in medicine may be thought of as the subspecialty of medicine that focuses on healing injuries and illnesses. Therapeutic treatments supported by scientific evidence are what we call "digital therapeutics" (DTx), which is a subset of digital medicine. By their definition, digital therapies "provide evidence-based therapeutic treatments to patients that are driven by high-quality software programs to prevent, manage, or treat a medical ailment or disease," as stated by the Digital Therapeutics Alliance. It entails employing a wide range of digital tools for patients' psychological and physiological health.

5 Problems and Challenges

5.1 *Holographic Simulation—Intelligent Processing*

Unity's new holographic simulation functionality drastically cuts down on development iteration time for holographic applications. Prototyping, debugging, and iterating designs can be done directly in the Unity Editor, saving time and effort for developers creating applications for Microsoft HoloLens. Despite the fact that the present results have little clinical significance.

5.2 Fusion of Virtual and Real—Quality Control

The advantages of the Internet of Things medical integration of virtual and real can be reflected in the AR glasses, allowing doctors to move away from the traditional mode of diagnosis and treatment in hospitals and clinics, and instead to ask virtual and real cloud experts to guide the diagnosis and treatment of end doctors at any time, from any location. Anywhere, anytime, researchers may monitor and direct pertinent research projects to ensure they meet national and international standards and increase conformity to the practices of contemporary assembly line engineering.

Effectively strengthening the seamless integration between participants (doctors, patients), the actual environment (devices), and the virtual environment is the primary goal of virtual-real fusion technology in metaverse medicine (virtual doctors, patients and devices). The end objective is to deliver medical services that involve the linking of humans and machines, achieving a level of natural fidelity between the two. Key technologies include high-precision positioning, virtual and real-world fusion presentation, optical display, and multi-sensory interaction are necessary to construct a virtual-real fusion environment. Concurrently, the issue of quality assurance using augmented reality glasses showing IoT intelligent diagnostic and management levels has to be addressed. In order to achieve this goal, it is necessary to build cutting-edge tools, do extensive planning, provide extensive training, and use cutting-edge quality control technology. The broad commercialization of AR glasses has not occurred yet, though. Therefore, implementing AR/VR systems in healthcare settings remains technically complex and prohibitively expensive, and the senior population frequently shows lower levels of acceptability for such technology. There is still a need for, and discussion on, a thorough guideline for standardizing these technologies in healthcare.

5.3 Virtual-Real Linkage—Human–Machine Integration

"Virtual-real linkage—human–machine integration" will be the clinical term that best reflects the medicinal significance of the metaverse. While the technology behind the virtual reality connection has advanced significantly in recent years, it still requires substantial human–machine integration to be useful in therapeutic settings. Therefore, it is suggested that a program, or "human–machine MDT," be developed to systematize the issue of "virtual-real linkage," "human–machine fusion," and so on [85]. Through the use of procedural digital technology, we can help transform the current model of diagnosis and treatment from a craft workshop into a state-of-the-art assembly line operation that can compete on a global scale. Theoretically, when people, computers, and the natural world work together, a new kind of intelligence emerges. It's not like regular people or robots; it has its own kind of intellect. It's a state-of-the-art scientific system that merges physical and biological elements. Human–computer integrated intelligence emphasizes brain-dominated intelligence mixed with "computer," whereas human–computer interaction technology focuses on physiological and psychological ergonomic difficulties. From a medical perspective, this partnership may be thought of as a "human–machine MDT," in which humans and machines work together to diagnose and treat patients. In the case of end-of-life care, for instance, a combination of virtual cloud specialists and end-of-life doctors can help standardize diagnosis and treatment according to consensus norms. In robotic surgery, for instance, doctors in the cloud remotely control surgical robots to treat patients. All aspects of health, from prevention to wellness to retirement planning, may benefit from this. Quality control according to national or international standards will be essential in the future of the metaverse in medicine, and APP can help with this. In the case of pulmonary nodule evaluation, for instance, a high degree of sensitivity and specificity in the diagnosis can be attained through the collaborative efforts of human experts and virtual, computerized artificial intelligence systems operating under the conditions of metaverse quality control. To do this, we must use not just deep learning to train the robot but also consensus rules and stringent quality control. The existing quality control is not all-time–space and automated, but Metaverse in Medicine may overcome these shortcomings, and utilize the simulated quality control robot to engage with quality control specialists to increase the overall quality control level with half the work.

5.4 Incompatibility

However, the implementation of this technique has its own set of challenges. Fortnite, Roblox, Horizon Worlds, and Microsoft Mesh, all current competitors to the Metaverse, are incompatible. Contrary to the original concept of the metaverse, there is currently no way to import or export material from Roblox to Horizon Worlds or vice versa. Historically, the doctor-patient interaction has been viewed as important to the practice of medicine. A doctor and patient have an initial consultation during which they address the presenting issue. The doctor then uses the patient's emotional and physical responses along with clinical data to diagnose the ailment. At last, the doctor makes a recommendation as to which therapeutic approach will work best for the patient. The advent of big data and artificial intelligence has radically altered modern life, but its implications are not always readily understood. Because of the primary focus on sight in VR and AR, the groundwork for their application in the delivery of medical services, which rely on all five senses, including touch, remains undeveloped. Developers that use terms like "virtual reality" and "metaverse" interchangeably further the confusion [86]. Eventually, the Metaverse worldview will supersede other service domains; that much is certain. Alteration and ease of use are also awaited with great eagerness. On the other side, our medical professionals should place a premium on treating patients with compassion while also prioritizing human dignity and respect for life. In order to keep people safe from the system, careful consideration must be given to all potential outcomes, including policies.

It's important to remember that for every opportunity there is in the health metaverse, there are still challenges. Patient confidentiality and life safety are two of the many issues that arise in the metaverse. A number of different aspects of the health metaverse will have profound effects on the way medicine is practiced in the future. Because of its vast user base and innovative linkages, the Metaverse inevitably gives rise to a variety of security challenges on a variety of scales. The health of individuals might be jeopardized if their efforts to maintain physical and mental wellness are not properly coordinated. Additionally, early Metaverse development should think about how to safeguard users' identities and well-being. It's fair to assume that the Metaverse, a network of interconnected computers, robots, and humans, will have serious security flaws, which begs the question of what kinds of oversight mechanisms will be put in place to maintain appropriate moral constraint. Health Metaverse's tech stack exemplifies the challenges of running an impenetrable network. The individualization of the doctor-patient interaction is jeopardized by these dangers. Interesting mental health problems are brought up by the existence of the metaverse, such as the potential for persons with psychosis, schizophrenia, depression, or anxiety to be injured, as well as the potential for the treatment of these conditions to evolve in general. The likes of Facebook, Microsoft, and others in the computer industry are presently leading the charge to popularize the concept of the "Metaverse". When the metaverse is ready, people will accept any form of censorship and will be easy targets for any number of corporations. According to research by Zhou et al., the economic model for the Metaverse is unfavorably skewed in favor of platform owners, which in turn hurts other rivals [87, 88].

6 Conclusion

Physical touch, eye contact, and the ability to read and respond to facial emotions and gestures are all crucial in the medical field, and these aspects cannot be replaced in a virtual environment. However, the metaverse may be seen as a tool to enhance the quality of health care intervention and treatment, global education, the assurance of uniform training, and research assistance in the creation of global databases. Experts have agreed on how to further improve the metaverse in medicine so that it can better assist the medical community and the healthcare system as a whole. Using the Cloud Plus Terminal platform, which incorporates augmented reality (AR) and virtual reality (VR) glasses and the medical Internet of Things, virtual and real cloud experts and terminal doctors were able to interact in the metaverse for the purposes of patient consultation, medical education, and the dissemination of scientific knowledge.

Medical services such as disease prevention, healthcare, physical examination, diagnosis, and treatment could all benefit from the use of the Metaverse, which has already been put to use in the development of diagnosis and treatment, and clinical researches. This is because the Metaverse could bring together virtual and real cloud experts and terminal users (such as terminal doctors, patients, and even their families) in a single environment.

In the field of medicine and healthcare, the Metaverse has both clinical and non-clinical use cases. Although trials have only been conducted in a small number of scenarios at present, we think that given the robust technological basis of the MIoT and the metaverse, it is only a matter of time before the metaverse is flawlessly used in all these scenarios. We can speed up our progress toward our goals if we adapt to the current situation and race against the clock. We can get closer to our goal of helping people by using the cutting-edge technologies of the metaverse in medical settings like these. Also, it's worth noting that a solid security infrastructure is crucial to the smooth running of the metaverse in medicine. Physical, system, operational, and managerial security can only be guaranteed if the entire security infrastructure is built with certain criteria in mind.

References

1. Bai, C.X.: Practical medical internet of things. People's Medical Publishing House, Beijing (2014)
2. Bai, C.X., Zhao, J.L.: Medical internet of things. Science Press, Beijing (2016)
3. Metaverse and healthcare: what's in it for you. Tech times—health biotech (2022). https://www.techtimes.com/articles/271630/20220209/metaverse-and-healthcare-what-s-in-it-for-you.htm. Accessed 12 July 2022
4. McKnight, R.R., Pean, C.A., Buck, J.S., Hwang, J.S., Hsu, J.R., Pierrie, S.N.: Virtual reality and augmented reality-translating surgical training into surgical technique. Curr. Rev. Musculoskelet. Med. **13**, 663–674 (2020)
5. Bernardo, A.: Virtual reality and simulation in neurosurgical training. World Neurosurg. **106**, 1015–1029 (2017)
6. New medical society studies how metaverse will change healthcare. Korea Biomed. Rev. (2022). http://www.koreabiomed.com/news/articleView.html?idxno=13049. Accessed 12 July 2022
7. Koo, H.: Training in lung cancer surgery through the metaverse, including extended reality, in the smart operating room of Seoul National University Bundang Hospital, Korea. J. Educ. Eval. Health Prof. **18**, 33 (2021)
8. Metaverse in Healthcare—New Era is Coming True. Healthcare Business Club (2022). https://healthcarebusinessclub.com/articles/healthcare-provider/technology/metaverse-in-healthcare/. Accessed 12 July 2022. The world's First Healthcare Metaverse from DeHealth. Cision; 2021. https://www.prnewswire.com/news-releases/the-worlds-first-healthcare-metaverse-from-dehealth-301449862.html. Accessed 12 July 2022
9. Maki, O., Alshaikhli, M., Gunduz, M., Naji, K.K., Abdulwahed, M.: Development of digitalization road map for healthcare facility management. IEEE Access **10**, 14 450–14 462 (2022)
10. Bhuiyan, M.N., Rahman, M.M., Billah, M.M., Saha, D.: Internet of things (iot): a review of its enabling technologies in healthcare applications, standards protocols, security, and market opportunities. IEEE Internet Things J. **8**(13), 10 474–10 498 (2021)
11. Kapoor, A., Guha, S., Das, M.K., Goswami, K.C., Yadav, R.: Digital healthcare: the only solution for better healthcare during covid-19 pandemic? 61–64 (2020)
12. Alshamrani, M.: Iot and artificial intelligence implementations for remote healthcare monitoring systems: a survey. J. King Saud Univ.-Comput. Inf. Sci. (2021)
13. Siriwardhana, Y., Gür, G., Ylianttila, M., Liyanage, M.: The role of 5g for digital healthcare against covid-19 pandemic: opportunities and challenges. ICT Express **7**(2), 244–252 (2021)

14. Shakeel, T., Habib, S., Boulila, W., Koubaa, A., Javed, A.R., Rizwan, M., Gadekallu, T.R., Sufiyan, M.: A survey on covid-19 impact in the healthcare domain: worldwide market implementation, applications, security and privacy issues, challenges and future prospects. Complex Intell. Syst. 1–32 (2022)
15. Park, S.-M., Kim, Y.-G.: A metaverse: taxonomy, components, ap- plications, and open challenges. IEEE Access **10**, 4209–4251 (2022)
16. Gadekallu, T.R., Huynh-The, T., Wang, W., Yenduri, G., Ranaweera, P., Pham, Q.-V., da Costa, D.B., Liyanage, M.: Blockchain for the metaverse: a review (2022). arXiv:2203.09738
17. Petrigna, L., Musumeci, G.: The metaverse: a new challenge for the healthcare system: a scoping review. J. Funct. Morphol. Kinesiol. **7**(3), 63 (2022)
18. Locurcio, L.: Dental education in the metaverse. Br. Dent. J. **232**(4), 191–191 (2022)
19. Zeng, Y., Zeng, L., Zhang, C., Cheng, A.S.: The metaverse in cancer care: applications and challenges. 100111 (2022)
20. Taheri, M., Kalnikaite, D.: A study of how virtual reality and brain computer interface can manipulate the brain. In: 2022 The 5th International Conference on Software Engineering and Information Management (ICSIM), pp. 6–10 (2022)
21. Wilson, T.: The metaverse and healthcare: opportunities, challenges, and tips for tech pioneers (2022). https://datafloq.com/read/metaverse-healthcare-opportunities-challenges-tips/
22. Garcia, L.M., Birckhead, B.J., Krishnamurthy, P., Mackey, I., Sackman, J., Salmasi, V., Louis, R., Maddox, T., Darnall, B.D.: Three-month follow-up results of a double-blind, randomized placebo-controlled trial of 8-week self-administered at-home behavioral skills-based virtual reality (vr) for chronic low back pain. J. Pain **23**(5), 822–840 (2022)
23. Elmer, N.A., Hassell, N., Comer, C.D., Bustos, V., Lin, S.J.: Plastic surgery in the metaverse. Plastic Surgery 22925503221109714. https://doi.org/10.1177/22925503221109714
24. Garavand, A., Aslani, N.: Metaverse phenomenon and its impact on health: a scoping review. Inf. Med. Unlocked 101029 (2022)
25. Huh, S.: Application of computer-based testing in the Korean medical licensing examination, the emergence of the metaverse in medical education, journal metrics and statistics, and appreciation to reviewers and volunteers. J. Educ. Eval. Health Prof. **19** (2022)
26. Xi, N., Chen, J., Gama, F., Riar, M., Hamari, J.: The challenges of entering the metaverse: an experiment on the effect of extended reality on workload. Inf. Syst. Front. 1–22 (2022)
27. Taylor, L., Dyer, T., Al-Azzawi, M., Smith, C., Nzeako, O., Shah, Z.: Extended reality anatomy undergraduate teaching: A literature review on an alternative method of learning. Annals of Anatomy-Anatomischer Anzeiger **239**, 151817 (2022)
28. Nakamatsu, N.A., Aytaç, G., Mikami, B., Thompson, J.D., Davis, M., III., Rettenmeier, C., Maziero, D., Stenger, V.A., Labrash, S., Lenze, S., et al.: Case-based radiological anatomy instruction using cadaveric mri imaging and delivered with extended reality web technology. Eur. J. Radiol. **146**, 110043 (2022)
29. JosephNg, P.S., Gong, X.: Technology behavior model—impact of extended reality on patient surgery. Appl. Sci. **12**(11), 5607 (2022)
30. Dadario, N.B., Quinoa, T., Khatri, D., Boockvar, J., Langer, D., D'Amico, R.S.: Examining the benefits of extended reality in neurosurgery: A systematic review. J. Clin. Neurosci. **94**, 41–53 (2021)
31. A. H. Kelly, J. Lezaun, I. Löwy, G. C. Matta, C. de Oliveira Nogueira, and E. T. Rabello, "Uncertainty in times of medical emergency: Knowledge gaps and structural ignorance during the brazilian zika crisis," Social Science & Medicine, vol. 246, p. 112787, 2020.
32. Faustino, S., Faria, I., Marques, R.: The myths and legends of king satoshi and the knights of blockchain. Journal of Cultural Economy **15**(1), 67–80 (2022)
33. Verde, F., Stanzione, A., Romeo, V., Cuocolo, R., Maurea, S., Brunetti, A.: Could blockchain technology empower patients, improve education, and boost research in radiology departments? an open question for future applications. J. Digit. Imaging **32**(6), 1112–1115 (2019)
34. Xiong, H., Jin, C., Alazab, M., Yeh, K.-H., Wang, H., Gadekallu, T.R., Wang, W., Su, C.: On the design of blockchain-based ecdsa with fault-tolerant batch verification protocol for blockchain-enabled iomt. IEEE J. Biomed. Health Inform. **26**(5), 1977–1986 (2021)

35. Bamakan, S.M.H., Nezhadsistani, N., Bodaghi, O., Qu, Q.: Patents and intellectual property assets as non-fungible tokens; key technologies and challenges. Sci. Rep. **12**(1), 1–13 (2022)
36. Nguyen, C.T., Hoang, D.T., Nguyen, D.N., Dutkiewicz, E.: Metachain: a novel blockchain-based framework for metaverse applications. In: 2022 IEEE 95th Vehicular Technology Conference:(VTC2022- Spring), pp. 1–5. IEEE (2022)
37. I. Yaqoob, K. Salah, R. Jayaraman, and Y. Al-Hammadi, "Blockchain for healthcare data management: opportunities, challenges, and future recommendations," Neural Computing and Applications, pp. 1–16, 2021.
38. I. Skalidis, O. Muller, and S. Fournier, "Cardioverse: The cardiovascular medicine in the era of metaverse," Trends in Cardiovascular Medicine, 2022.
39. Monrat, A.A., Schelén, O., Andersson, K.: A survey of blockchain from the perspectives of applications, challenges, and opportunities. IEEE Access **7**, 117 134–117 151, 2019.
40. Gupta, R., Srivastava, D., Sahu, M., Tiwari, S., Ambasta, R.K., Kumar, P.: Artificial intelligence to deep learning: machine intelligence approach for drug discovery. Mol. Diversity **25**(3), 1315–1360 (2021)
41. Anter A.M., Moemen, Y.S., Darwish, A., Hassanien, A.E.: Multi-target QSAR modelling of chemo-genomic data analysis based on extreme learning machine. J. Knowl.-Based Syst. Elsevier, Knowl.-Based Syst. **188**, 104977 (2020)https://doi.org/10.1016/j.knosys.2019.104977
42. Dlamini, Z., Francies, F.Z., Hull, R., Marima, R.: Artificial intelligence (ai) and big data in cancer and precision oncology. Comput. Struct. Biotechnol. J. **18**, 2300–2311 (2020)
43. Huynh-The, T., Pham, Q.-V., Pham, X.-Q., Nguyen, T.T., Han, Z., Kim, D.-S., Artificial intelligence for the metaverse: a survey (2022). arXiv:2202.10336
44. Wu, E., Wu, K., Daneshjou, R., Ouyang, D., Ho, D.E., Zou, J.: How medical ai devices are evaluated: limitations and recommendations from an analysis of fda approvals. Nat. Med. **27**(4), 582–584 (2021)
45. Srivastava, G., Jhaveri, R.H., Bhattacharya, S., Pandya, S., Maddikunta, P.K.R., Yenduri, G., Hall, J.G., Alazab, M., Gadekallu, T.R., et al.: Xai for cybersecurity: state of the art, challenges, open issues and future directions (2022). arXiv:2206.03585
46. Wang, S., Qureshi, M.A., Miralles-Pechuaán, L., Huynh-The, T., Gadekallu, T.R., Liyanage, M.: Explainable ai for b5g/6g: Technical aspects, use cases, and research challenges (2021). arXiv:2112.04698
47. Munirathinam, S.: Industry 4.0: industrial internet of things (iiot). Adv. Comput. Elsevier **117**(1), 129–164 (2020)
48. Akbar, A., Kousiouris, G., Pervaiz, H., Sancho, J., Ta-Shma, P., Carrez, F., Moessner, K.: Real-time probabilistic data fusion for large-scale iot applications. IEEE Access **6**, 10 015–10 027 (2018)
49. Tun, S.Y.Y., Madanian, S., Mirza, F.: Internet of things (iot) applications for elderly care: a reflective review. Aging Clin. Exp. Res. **33**(4), 855–867 (2021)
50. Sharma, P.K., Ghosh, U., Cai, L., He, J.: Guest editorial: security, privacy, and trust analysis and service management for intelligent internet of things healthcare. IEEE Trans. Ind. Inf. **18**(3), 1968–1970 (2021)
51. Farahat, M.A., Abdo, A., Kassim, S.K.: A hybrid approach for protecting clinical and genomic data using lossless Stego-DNA compression. In: Silhavy, R., Silhavy, P., Prokopova, Z. (eds.) Software Engineering Application in Informatics. CoMeSySo 2021. Lecture Notes in Networks and Systems, vol. 232. Springer, Cham (2021). https://doi.org/10.1007/978-3-030-90318-3_48
52. Farahat, M.A., Abdo, A., Kassim, S.K.: A systematic literature review of DNA-based steganography techniques: research trends, data sets, methods, and frameworks. In: Magdi, D.A., Helmy, Y.K., Mamdouh, M., Joshi, A. (eds.) Digital Transformation Technology. Lecture Notes in Networks and Systems, vol. 224. Springer, Singapore (2022). https://doi.org/10.1007/978-981-16-2275-5_31
53. Pham, Q.-V., Fang, F., Ha, V.N., Piran, M.J., Le, M., Le, L.B., Hwang, W.-J., Ding, Z.: A survey of multi-access edge computing in 5g and beyond: Fundamentals, technology integration, and state-of-the-art. IEEE Access **8**, 116 974–117 017 (2020)

54. Siriwardhana, Y., Porambage, P., Liyanage, M., Ylianttila, M.: Ai and 6g security: opportunities and challenges. In: 2021 Joint European Conference on Networks and Communications & 6G Summit (EuCNC/6G Summit), pp. 616–621. IEEE (2021)
55. Dogra, A., Jha, R.K., Jain, S.: A survey on beyond 5g network with the advent of 6g: architecture and emerging technologies. IEEE Access **9**, 67 512–67 547 (2020)
56. Sun, M., Xie, L., Liu, Y., Li, K., Jiang, B., Lu, Y., Yang, Y., Yu, H., Song, Y., Bai, C., et al.: The metaverse in current digital medicine. Clinical eHealth (2022)
57. Moglia, A., Georgiou, K., Marinov, B., Georgiou, E., Berchiolli, R.N., Satava, R.M., Cuschieri, A.: 5g in healthcare: from covid-19 to future challenges. IEEE J. Biomed. Health Inform. **26**(8), 4187–4196 (2022)
58. Ramu, S.P., Boopalan, P., Pham, Q.-V., Maddikunta, P.K.R., Huynh-The, T., Alazab, M., Nguyen, T.T., Gadekallu, T.R.: Federated learning enabled digital twins for smart cities: concepts, recent advances, and future directions. Sustain. Cities Soc. **79**, 103663 (2022)
59. Moyne, J., Qamsane, Y., Balta, E.C., Kovalenko, I., Faris, J., Barton, K., Tilbury, D. M.: A requirements driven digital twin framework: Specification and opportunities. IEEE Access **8**, 107 781–107 801 (2020)
60. Madni, A.M., Madni, C.C., Lucero, S.D.: Leveraging digital twin technology in model-based systems engineering. Systems **7**(1), 7 (2019)
61. Gehrmann, C., Gunnarsson, M.: A digital twin based industrial automation and control system security architecture. IEEE Trans. Ind. Inf. **16**(1), 669–680 (2019)
62. Han, Y., Niyato, D., Leung, C., Kim, D.I., Zhu, K., Feng, S., Shen, S.X., Miao, C.: A dynamic hierarchical framework for iot-assisted digital twin synchronization in the metaverse. IEEE Internet Things J. (2022)
63. Möller, J., Pörtner, R.: Digital twins for tissue culture techniques—concepts, expectations, and state of the art. Processes **9**(3), 447 (2021)
64. De Mauro, A., Greco, M., Grimaldi, M.: A formal definition of big data based on its essential features. Library Rev. (2016)
65. Karatas, M., Eriskin, L., Deveci, M., Pamucar, D., Garg, H.: Big data for healthcare industry 4.0: applications, challenges and future perspectives. Expert Syst. Appl. 116912 (2022)
66. Wu, J., Gao, G.: Edu-metaverse: Internet education form with fusion of virtual and reality. In: 2022 8th International Conference on Humanities and Social Science Research (ICHSSR 2022), pp. 1082–1085. Atlantis Press (2022)
67. Depari, G.S., Shu, E., Indra, I.: Big data and metaverse toward business operations in indonesia. Jurnal Ekonomi **11**(01), 285–291 (2022)
68. Hidary, J.D., Hidary, J.D.: Quantum Computing: An Applied Approach, vol. 1. Springer (2021)
69. Ajagekar, A., You, F.: Quantum computing and quantum artificial intelligence for renewable and sustainable energy: a emerging prospect towards climate neutrality. Renew. Sustain. Energy Rev. **165**, 112493 (2022)
70. Ajagekar, A., You, F.: New frontiers of quantum computing in chemical engineering. Korean J. Chem. Eng. 1–10 (2022)
71. Alnuaim, A.A., Zakariah, M., Shukla, P.K., Alhadlaq, A., Hatamleh, W.A., Tarazi, H., Sureshbabu, R., Ratna, R.: Human-computer interaction for recognizing speech emotions using multilayer perceptron classifier. J. Healthc. Eng. **2022** (2022)
72. Liu, Y., Sivaparthipan, C., Shankar, A.: Human–computer interaction based visual feedback system for augmentative and alternative communication. Int. J. Speech Technol. **25**(2), 305–314 (2022)
73. Feng, X., Jiang, Y., Yang, X., Du, M., Li, X.: Computer vision algo- rithms and hardware implementations: a survey. Integration **69**, 309–320 (2019)
74. Cao, K., Liu, Y., Meng, G., Sun, Q.: An overview on edge computing research. IEEE Access **8**, 85 714–85 728 (2020)
75. Prabadevi, B., Deepa, N., Pham, Q.-V., Nguyen, D.C., Reddy, T., Pathirana, P.N., Dobre, O., et al.: Toward blockchain for edge-of-things: a new paradigm, opportunities, and future directions. IEEE Internet Things Mag. **4**(2), 102–108 (2021)

76. Dhelim, S., Kechadi, T., Chen, L., Aung, N., Ning, H., Atzori, L.: Edge-enabled metaverse: The convergence of metaverse and mobile edge computing. arXiv:2205.02764 (2022)
77. Jiang, X., Yu, F.R., Song, T., Leung, V.C.: A survey on multi-access edge computing applied to video streaming: some research issues and challenges. IEEE Commun. Surv. & Tutor. **23**(2), 871–903 (2021)
78. Shah, P., Luximon, Y.: Three-dimensional human head modelling: a systematic review. Theor. Issues Ergon. Sci. **19**(6), 658–672 (2018)
79. Gao, Q., Gadekallu, T.R.: Design of telemedicine information query system based on wireless sensor network. EAI Endorsed Trans. Pervasive Health Technol. **8**(4), e1–e1 (2022)
80. Taiwo, O., Ezugwu, A.E.: Smart healthcare support for remote patient monitoring during covid-19 quarantine. Inf. Med. Unlocked **20**, 100428 (2020)
81. Jnr, B.A.: Use of telemedicine and virtual care for remote treatment in response to covid-19 pandemic. J. Med. Syst. **44**(7), 1–9 (2020)
82. Fateel, E.E., O'Neill, C.S.: Family members' involvement in the care of critically ill patients in two intensive care units in an acute hospital in bahrain: The experiences and perspectives of family members' and nurses'-a qualitative study. Clin. Nurs. Stud. **4**(1), 57–69 (2016)
83. Keller, D.S., Grossman, R.C., Winter, D.C.: Choosing the new nor- mal for surgical education using alternative platforms. Surg. Infect. (Larchmt.) **38**(10), 617–622 (2020)
84. Gelernter, D., Mirror worlds: Or the day software puts the universe in a shoebox... How it will happen and what it will mean. Oxford University Press (1993)
85. Yang, K., Li, Z., Yuan, Y.F., Chen, Z.Q., Lei, H., Wang, X.H.: Application of stepped training for the surgical robots. China Higher Med. Educ. **3** (2018)
86. Chen, D., Zhang, R.: Exploring research trends of emerging technologies in health metaverse: a bibliometric analysis. SSRN Electron J. (2022). https://doi.org/10.2139/ssrn.3998068
87. Lee RA: Is it possible to apply the Metaverse to the health care system? Ewha Med. J. **45**, 12 (2022). https://doi.org/10.12771/emj.2022.45.1.1
88. Metaverse in Healthcare: Potential to be the Next Frontier? (2022). https://www.informationweek.com/big-data/metaverse-the-next-frontier-in-healthcare. Accessed 12 July 2022

Emerging Technologies in Metaverse World

A Proposed Metaverse Framework for Food Security Based-IoT Network and Machine Learning

Lobna M. Abou El-Magd, Noha M. M. Abdelnapi, Ashraf Darwish, and Aboul Ella Hassanien

Abstract Metaverse is a digital environment that enables user interaction. It is evolving into a parallel universe in which people can work, play, and interacts. Using mixed reality (MR) technology, Metaverse integrates technologies with ambient intelligence to establish a bridge between the digital and physical realities, allowing users to combine resources. When IoT is enabled, the Metaverse can build a fully integrated partnership with the physical world. In order to integrate digital, IoT plays a crucial role in the Metaverse by developing an interoperable, seamless system of systems. Meat is a crucial source of protein for the human body; however, it is susceptible to spoilage, which hurts human health. As a result, in this chapter, we use meta-veterinary in the realm of food security to detect spoiled meat. The proposed approach collects data from the physical environment using IoT gas sensors and then applies a machine-learning algorithm to judge meat quality. The Support Vector Machine (SVM) algorithm is utilized in the classification model for meat quality, and it has three different kernels: linear, cubic, and quadratic. A linear accuracy rate

L. M. Abou El-Magd (✉)
Computer Science Department, MET High Institute, Mansoura, Egypt
e-mail: lobna_aboelmagd@metmans.edu.eg
URL: http://www.egyptscience.net

N. M. M. Abdelnapi
Faculty of Computers and Information, Suez University, Suez, Egypt
e-mail: Noha.abdelnapi@fci.suezuni.edu.eg
URL: http://www.egyptscience.net

A. Darwish
Faculty of Science, Helwan University, Cairo, Egypt
e-mail: ashraf.darwish.eg@ieee.org
URL: http://www.egyptscience.net

A. E. Hassanien
Faculty of Computers and AI, Cairo University, Cairo, Egypt
e-mail: aboitcairo@cu.edu.eg
URL: http://www.egyptscience.net

L. M. Abou El-Magd · N. M. M. Abdelnapi · A. Darwish · A. E. Hassanien
Scientific Research Group in Eqypt, Giza, Egypt

A. E. Hassanien et al. (eds.), *The Future of Metaverse in the Virtual Era and Physical World*, Studies in Big Data 123, https://doi.org/10.1007/978-3-031-29132-6_8

of 97.4%, quadratic accuracy rate of 99.4%, and cubic accuracy rate of 99.9% were achieved by the performance.

Keywords Metaverse · IoT · Machine learning · Virtual reality · Augmented reality · Mixed reality · Food security

1 Introduction

The Metaverse is defined as a place where the real world and the digital world meet. Imagine a future where everything you need, from your closest friends and family to your favorite stores and restaurants to a map of the world and the cosmos itself, exists solely in the digital realm. The term "Metaverse," a combination of "Meta" and "Universe," was coined in 1992. It is a 3D virtual world that tries to simulate the real world as closely as possible. Several new companies (like Decentraland, Sandbox, Upland, etc.) and many well-known IT companies (like Facebook or Meta, Microsoft, Google, Samsung, etc.) are interested in the Metaverse and are trying to make their Metaverses available as new services [1].

You can think of the Metaverse as a better version of the Internet. In reality, the Metaverse doesn't depend on a computer screen. Instead, it provides the perfect way for people to have large-scale social interactions. People can talk to each other while playing games, watching movies, or virtually going to a popular tourist spot. Metaverse is able to make the internet more useful by using 3D technologies, Virtual Reality (VR), Augmented Reality (AR), and Artificial Intelligence (AI) [2–5].

The Internet of Things (IoT) can be employed in the Metaverse, which mappings real-time IoT data from the physical world into a digital reality in the virtual world. Because of this, wireless, connected, and immersive digital experiences are now possible [6]. IoT can add to how AR/VR lets users interact with the virtual world [7].

Despite the fact that IoT already connects real items to the Internet. However, integrating Metaverse can unquestionably provide customers with new opportunities. These technologies, when combined, would enable the transfer of real-time data from the physical world to the digital realm. Thus, the gap between the real and virtual worlds would be significantly reduced. IoT devices will be more online and offline connected with the help of the Metaverse. Additionally, customers might easily construct their own digital avatars using IoT devices.

Food safety is a huge issue on a global scale, both in terms of human suffering and economic expenses. Scientific breakthroughs have expanded our understanding about the nutritional qualities of foods and their implications on health. This indicates that a big majority of customers are far more aware of what they eat and have higher standards for food quality. In addition to safety, food quality is a broad phrase that includes inherent features such as appearance, colour, texture, and flavour, as well as external characteristics such as perception or involvement [8].

The meat sector faces a serious problem with food-borne illnesses. For the industry and sanitary management, research on pathogens found in meat and meat products is essential. One of the most significant and well researched pathogens found in fresh meat is Escherichia coli, which has been recognized as a contamination of raw, non-intact beef products since 1999. To lessen this risk, some strategies should be put into practice. The introduction of new infections is a worry as well. Due to their widespread incidence and prominence as leading causes of human gastroenteritis, Salmonella spp., Listeria monocytogenes, enterotoxigenic, Staphylococcus aureus, and Campylobacter spp. constitute this category of pathogenic bacteria's principal food contaminants Achieving food security for all is a top priority for sustainable development. The protein in meat is beneficial to human health because it helps repair and maintain tissues. It was crucial to monitor meat quality because it is a perishable item.

In this chapter, the Metaverse is used to assess the quality of the meat through the use of IoT devices and a machine learning algorithm in the role of a veterinary avatar.

This chapter is structured as follows. In Sect. 2, the related works are highlighted. Section 3 presents the basics and background. Section 4 presents the proposed framework. Section 5 presents the results of the experiments and analyses the results and performance of the proposed model for classifying the quality of meat. Finally, Sect. 6 presents the conclusion and future work.

2 Related Work

In this part, we examine recent research on IoT applications in the Metaverse. The evolution of virtual worlds over several decades, from text-based settings to the Metaverse, is outlined in Dionisio et al. [9]. The survey summarizes four features according to physical and virtual interaction: realism (in a way that enhances the game for the player), ubiquity (letting existing IoT devices have full access to the system while keeping the user's virtual identity across different apps.), interoperability (allowing the user to enjoy an unbroken seamless virtual experience), and scalability (enabling a high number of concurrent users to run Metaverse applications).

In Lee et al. [10], a digital twins-native continuum is outlined, outlining the evolution of the Metaverse in three stages: digital twins of humans and IoT, the creation of native content, and the merger of actual IoT devices with virtual worlds.

To solve the problem of synchronization between Digital Twins (DT) for VSPs in the Metaverse with the help of IoT devices, Han et al. [11] proposed a dynamic hierarchical architecture. In particular, they advocated for value decay dynamics in time to quantify DT values and their impact on VSP synchronization. VSPs can also benefit from a slew of IoT sensors that provide real-time updates on the health of the real-world devices standing in for DTs. Then, they used an evolutionary game to model the sensor-picking habits of the evolving VSPs.

The research presented in Gao [12] investigated the new situation of IoT-supported Metaverse marketing in an effort to increase the quality of human connections and the communication efficiency of marketing activities. The results demonstrated that, when applied to the field of business marketing, the customer manager will no longer be limited to simple voice and image, but will also be able to perceive the signals emitted by the customer's micro expressions and body language to achieve a more effective interaction with consumers.

Metaverse applications confront intrinsic disconnects between virtual and physical IoT components and interfaces. According to Guan et al. [13], a Metaverse framework can be utilized to improve the seamless integration of interoperable agents between virtual and physical environments.

Immersion, diversity, economy, politeness, interaction, authenticity, and independence are the most essential prerequisites for constructing the IoT-inspired Metaverse. The authors of Li et al. [14] investigated six popular Internet of Things applications found in the Metaverse. These applications included collaborative entertainment, smart city, education, healthcare, real estate, and social networking. Seong-Soo [15] looked into the future of the food industry in terms of Non-Fungible Tokens (NFTs) and the Metaverse in the food industry. They also looked at the ecosystems of the retail and food and beverage industries to see how the Metaverse could be used in the future. They then made suggestions about how the Metaverse's future could change things.

Nam [16] discussed BGF Retail's CU convenience store building a store in ZEPETO to provide food services linked to the actual store, and the luxury fashion brand GUCCI creating ads and branded objects on the platform to make money. It's great that the entertainment industry makes celebrity-fan interaction easier. For instance, fans can chat with celebrity avatars and buy celebrity-themed food. XR, a Metaverse technology, is gaining popularity. XR uses lifelike remote communication.

3 Basics and Background

3.1 Metaverse

The purpose of the Metaverse is to develop an interconnected network of 3D virtual and real-world environments. This network will expand a single, global Internet to provide users with realistic cyber-virtual experiences that take place in real-world environments. Augmented reality (AR) and virtual reality (VR) are two well-known applications that have been developed to offer users of the Metaverse connected, immersive digital experiences as well as opportunities for social interaction [17].

3.2 The Metaverse's Virtual and Mixed Reality

Virtual reality, augmented reality, and haptic Internet are examples of immersive technology. The term "immersive technology" refers to tools that allow users to have a "whole sensory immersion" in a given environment, which may include the use of multiple senses, the manipulation of virtual objects, and the blending of real and virtual environments.

Virtual reality (VR) aims to create a new type of human–machine interaction in which users can have an immersive, first-person experience in a computer-generated, three-dimensional environment. Through the use of virtual reality technology, users can interact with a simulated environment and carry out a variety of tasks. VR allows a user to completely submerge oneself in a simulated setting, allowing them to experiment with their senses and develop novel modes of interaction and understanding. In mixed reality (MR), real-world and virtual elements coexist and interact in real time. Blending digitally into reality with MR, either virtual items are added to the actual world or real world objects are added to the virtual world. In the latter case, which is sometimes referred to as augmented virtuality, real-world elements in a virtual world take on a sense of permanence and appear to actually exist within the virtual world, which is a step closer to a VR environment [18].

In a virtual environment, physical objects can be controlled digitally, giving the impression of permanency. However, MR may also make it easier to interact with digital elements that are fixedly affixed to the real world. No matter the circumstances, teleoperations will always include MR, and vice versa.

3.3 *Metaverse Architecture*

A general Metaverse architecture consisting of three layers was proposed by Duan et al. in the year 2021 [19]. The Interaction layer serves as a link between the Ecosystem layer and the Infrastructure layer in this suggested broad architecture of the Metaverse. The seven layers of the Metaverse that were stated are broken down into the following three phases, according to the architecture that was provided by Duan:

(1) Infrastructure: This layer is responsible for the establishment of fundamental and physical necessities, such as the blockchain, the network, and the computational capacities.
(2) Interaction: This layer serves as a connection between the Ecosystem and Infrastructure layers, and it is also where the contents of the Metaverse are formed.
(3) Ecosystem: This refers to the alternative digital world, often known as the Metaverse. This layer incorporates economics, artificial intelligence, and user-generated content.

3.4 Internet of Things

In the context of the IoT, every connected device is seen as a thing. Most things have physical actuators, sensors, and an embedded system with a microprocessor. Machine-to-Machine (M2M) communication is needed because things need to talk to each other. Short-range wireless communication can be done with Wi-Fi, Bluetooth, and ZigBee. Long-range wireless communication can be done with mobile networks like WiMAX, LoRa, Sigfox, CAT M1, NB-IoT, GSM, GPRS, 3G, 4G, LTE, and 5G [19]. By collecting and analysing huge amounts of data, IoT can help make many processes more measurable and quantifiable [20].

IoT has the potential to improve the quality of life in many areas, such as healthcare, smart cities, the construction industry, agriculture, water management, and the energy sector [21]. It is used a lot in environmental monitoring, healthcare systems and services, energy-efficient building management, and drone-based service delivery [22, 23].

The following are a few ways Metaverse will impact IoT:

I. Real-Life Training

Metaverse relies on AI and experience. Metaverse in IoT can improve training by using virtual simulations. IoT can help acquire more accurate and timely real-world data for exam preparation. Metaverse participants can develop more advanced software or AI algorithms to discover flaws and show real-world effects. Thus, the Internet of Things makes the Metaverse's virtual environment accurate.

II. User Interface Personalization

IoT devices interact with other devices and feature simple interfaces. Metaverse can give typical IoT devices with screens a 3D digital user experience. Users will have a more immersive experience utilizing IoT devices. Real-world and virtual presence will be possible. Thus, firms can hire IoT app developers to customize user interfaces and experiences. When IoT is deployed, the Metaverse will feel more like the physical world. People, IoT devices, and the Metaverse's complicated environment and processes will interact more. Due of the Metaverse's immersiveness and real-world use cases; we'll make better decisions with less learning and training.

III. Long-Term Planning Effective

The Metaverse constantly adds digital content based on real-world objects including buildings, people, automobiles, clothing, etc. Businesses try to recreate our physical environment in internet. With IoT and Metaverse, businesses may plan for numerous scenarios and better their long-term goals. Real-time data is needed for realistic scenario simulation and long-term planning. Energy, transportation, healthcare, fashion, and other sectors may profit. Most importantly, AI and machine learning may impact long-term planning.

IV. Cloud Technology's Significant Impact

Metaverse and IoT depend on cloud technologies. However, their combination could transform cloud technology. The Metaverse's incorporation into IoT devices may change cloud platforms like Amazon AWS and Microsoft Azure. Cloud service providers might unlock cloud computing's full potential with the IoT Metaverse. Cloud technology developers can use structural frameworks to improve scalability, robustness, and seamless functionality. Cloud computing users will benefit from the Metaverse's processing capacity.

V. Interactive Computing

Most computer interactions happen on a screen or input device. IoT devices require console or button input. AR and VR will change IoT device interaction in the Metaverse. Users won't even know if they're in real or virtual space. It will be realistic and contextualized. Extended reality IoT gadgets will enable this. IoT with motion detection, AI-enabled edges, and customized data-gathering sensors will enable immersive, interactive computing.

4 The Proposed Metaverse Framework for Food Security

The proposed framework presents of three main parts; virtual world, Metaverse engine, and physical world. Next we will present these parts in details (Fig. 1).

4.1 Virtual World

I. **Avatars**

Digital avatars can simulate human behavior in a Metaverse-based system. Since humans immediately interact with other mobile gadgets like autonomous vehicles (AVs) and UAVs, avatars (digital copies) should also reflect them. However, roads, traffic signs, and buildings should be represented in the digital twin model. Thus, we must use digital twins, avatars, and Metaverse interactive experience technologies to represent wireless system physicality. Consider automobile and AV accident reporting. Digital twins can represent environment objects like RSUs. If an actual collision occurs, the Metaverse will show how people, AVs, and other automobiles react, generating anomalous and hotspot wireless traffic and mobility patterns. Thus, Metaverse interactive experience technologies can add information to twin models and avatars in meta space. Machine learning modules and human network administrators can analyze/control physical entities like UAV flying base stations and edge

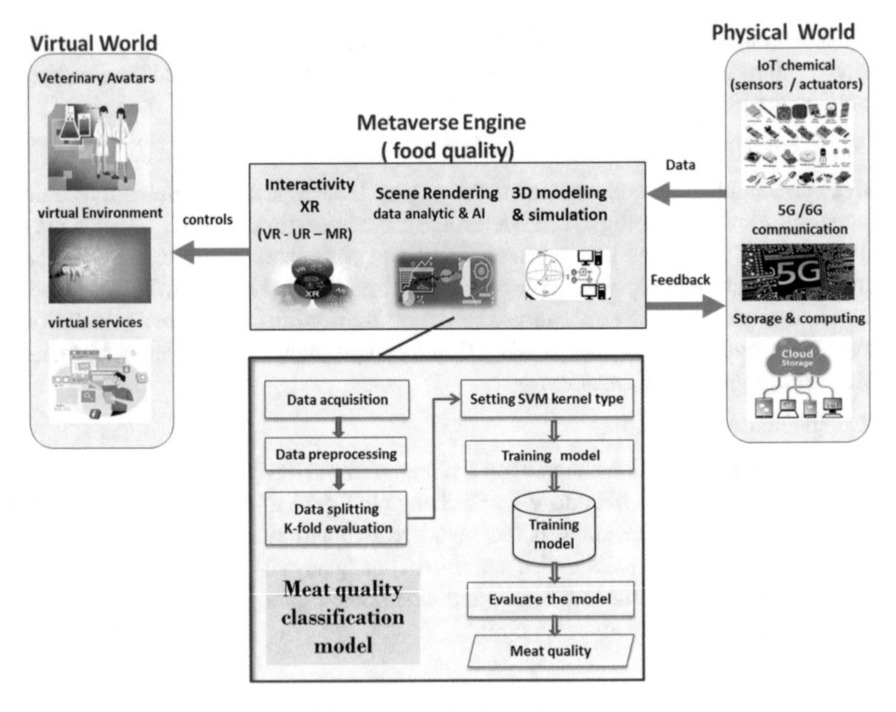

Fig. 1 The proposed Metaverse framework for food security

servers to report accidents and other functions (e.g., lane change assistance and collision avoidance) [24].

In the suggested framework, the digital avatar is initially configured by a veterinarian.

II. The Virtual environment

From tools to virtual furniture, everything is included. We can divide these into events, settings, and persons. The "scene" in our work is the indoor setting, medical equipment, and the "veterinarian." Dynamic data and information from viewer-viewer interactions make up events [25].

To make the virtual environment more realistic, a dynamic meta-process will be established. On the other side, technology will improve physical loyalty. It will also be linked to meat sensor data. Veterinarians will receive real-time, intuitive feedback from digital avatars.

4.2 Metaverse Engine

I. **Interactivity**

The Metaverse will provide virtual worlds with AR, VR, extended reality (XR), and MR as the primary technologies that enable interactions between virtual models (i.e., digital twins) and avatars (i.e., human models) (MR).

The technology blurs the distinction between the virtual and physical worlds by using virtual reality, augmented reality, and artificial intelligence. Put on your special accessories, such as smart glasses, VR-AR compatible headsets, and remote controls, and you'll soon find yourself submerged in the Metaverse world. In his best-selling book Snow Crash, Neal Stephenson introduced the concept of virtual world space [26].

II. **Scene rendering**

A crucial step in giving virtual objects their true features and putting them in front of our eyes is rendering. Lighting, texturing, and colouring are typically included in rendering. Since laser scanning lacks colour information that is required in many applications. In order to colourize the point cloud data, a hybrid 3D reconstruction based on scan data and photos is used. This method enables the development of more lifelike "virtual items" in the Metaverse by significantly increasing sampling efficiency and more precisely restoring material attributes [27]. In the architecture we've suggested, we employ a machine learning algorithm to decide whether to draw a scene based on the output classification. It will be presented in the part after this.

III. **Meat quality classification model based on SVM**

A supervised statistical learning method, support vector machine (SVM) is widely used in the field of machine learning. By minimizing structural risk, SVM looks for an optimal hyper-plane in a kernel space where training cases are linearly separable. Typically, SVM is processed using the so-called soft-margin. SVM Kernel approaches are advantageous for the effective implementation of SVM. Kernel methods are a collection of techniques for mapping data from the original feature space to the kernel space without knowing the explicit mapping function Φ. The following is how kernel functions define inner product spaces (Hilbert spaces):

$$K(X_i, X_j) = \langle \Phi(X_i), \Phi(X_j) \rangle \tag{1}$$

SVM is able to handle both linear and nonlinear issues since it is powered by a good kernel. In kernel-based approaches, such as SVM, the kernel function is the most essential component. According to Hilbert–Schmidt theory and the Mercer condition, several kernel functions are generated [28].

During the SVM-based categorization of meat quality, three distinct Kernel functions are utilized: linear, quadritic, and cubic kernels.

IV. **3D Modeling and Simulation**

The graphical methods used to produce the integrated environment fusing the real and virtual worlds, such as the 3D development of scenery, non-player characters (NPCs), and player characters, provide the basis for Metaverse's visual creation (Avatar). Users can freely explore the Metaverse, interact with visual components, and have an immersive experience thanks to interactive technologies. Additional guidelines and recommendations are required to improve users' comprehension of the virtual world. Visualization can provide such guidance by processing Metaverse data and displaying it in an appropriate format to users. The evolution of these technologies makes the Metaverse more lifelike and engaging for users to perceive and explore [27]. As computer simulation technology continues to advance, digital technology will make the virtual medical environment appear more real. The distinction between the virtual and real worlds will become more hazy when a mixed medical reality (MR) environment is created. The user can, for instance, communicate with a physician who is using XR technology. Real-time data and information pertaining to meat will be extracted and shown as an augmented reality, establishing a new link with the existing natural world.

4.3 Physical Worlds

I. **IoT (sensors and actuators)**

The Internet of Things will be vital in addressing one of the most severe threats lurking over the Metaverse area. That is, the capacity of the Metaverse to gather information from the real world and introduce it into the virtual one. The data mapped from the real world must be precise, planned, secure, pertinent, and real-time in order to produce the desired results. IoT is a fantastic opportunity for the Metaverse to get noticed because it has been there for a while and has thousands of cameras, sensors, and tech gadgets in its cluster.

In this work, there are number of gas sensors that are used to collect smell data of the meat such as MQ123 sensor which measures CO_2, NH_3, alcohol, ben-zene and smoke, this sensor and the other gases sensors are listed in Table 1.

We use the dataset published in Rahman Wijaya et al. [29], the dataset consists of five recorded time series, each of which corresponds to one of five beef cuts and has 2160 min of measurement points. The data contains; the reading values for the sensors listed in Table 1, time of measurement point (minutes) and Class label of beef quality ('excellent', 'good', 'acceptable', 'spoiled').

We've gathered the five portions of the described dataset, each of which has 2,160 rows, totaling 10,800 rows. Table 1 shows the measurements of the 10 sensors, and Table 2 shows the category's class names with the number of each category.

Table 1 List of gas sensors

No.	Gas sensor	Selectivity
1	MQ135	Carbon dioxide (CO_2), ammonia (NH_3), Nor, alcohol, ben-zene, smoke
2	MQ136	Hydrogen sulfide (H_2S)
3	MQ2	Liquefied petroleum gas (LPG), I-butane, propane, methane, alcohol, smoke
4	MQ3	Methane (CH_4), hexane, LPG, CO, alcohol, benzene
5	MQ4	Methane (CH_4), natural gas y
6	MQ5	LPG, natural gas, town gas
7	MQ6	Propane, LPG, iso-butane
8	MQ8	Hydrogen (H_2)
9	9MQ9	Propane, methane, CO
10	DHT22	Temperature, humidity

Table 2 Classes of the dataset

Class name	Number of class members
Excellent	1680
Good	2820
Acceptable	2100
Spoiled	4200
Total	10,800

II. 5G/6G Communication

The fifth generation of mobile communication technologies (5G) is needed because the amount of data traffic, the number of devices, and the types of service scenarios are all growing at a very fast rate. In the 5G era, mobile networks will not only serve mobile phones, but also tablets, mobile vehicles, and different types of sensors [30].

The sixth generation of mobile communication technologies (6G) supports total network coverage, better intelligent applications, and stronger security for a fully connected and smart digital world. 6G mobile communication technology will go beyond individualized communication to create intelligent connections between people, devices, and resources, moving from "connected things" to "connected intelligence." AR/VR, high-fidelity XR/MR, and even holographic communication will be produced in the 6G era. This crucial transition relies on XR fully mobilizing the senses of sight, touch, hearing, and smell to allow users to enjoy fully immersive holographic experiences like virtual sports, travel, and games whenever and everywhere. Based on 5G autopilot, the 6G era will enable on-demand, point-to-point smart travel [31].

III. **Storage & Computing**

The Internet of Things and Metaverse integration is founded on cloud technologies. Cloud service providers will foster innovation that unlocks advanced industry capacities. Scalability, robustness, and seamless functionality will be the defining characteristics of the infrastructures developed by developers. Popular cloud platforms such as Amazon Web Services (AWS) and Microsoft enable developers and organizations to leverage the greatest data services at their convenience and discretion.

Lack of reliable cloud services would obstruct data flow, which over time would harm cluster participants. This will have an effect on the amount of processing power necessary to send the data gathered by IoT devices to the Metaverse cluster in real time. It will be possible to create creative solutions to real-world issues thanks to improved interoperability between AR-VR devices and IoT data, supported by cloud providers.

5 Experiments, Results, and Discussion

Tensor flow and Keras in a Google Colab environment were used to conduct the experiments. In the classification problem, four evaluation indexes are used to evaluate the proposed approach's prediction ability: accuracy, precision, recall, and F1-score. Accuracy is the ratio of correct forecasts to all predictions, usually expressed as a percentage and determined using Eq. (2). Precision is a metric that assesses a model's ability to correctly forecast values in a given category and is calculated using Eqs. (3). The fraction of successfully recognized positive patterns is measured by recall, which is determined using Eq. (4). As seen in Eq. (5), the F1-score is the weighted average of precision and recall.

$$\text{Accuracy} = (\text{TP count} + \text{TN count})/(\text{TP count} + \text{FP count} + \text{FN count} + \text{TN count}) \tag{2}$$

$$\text{Precision} = \text{TP count}/(\text{TP count} + \text{FP count}) \tag{3}$$

$$\text{Recall} = \text{TP count}/(\text{TP count} + \text{FN count}) \tag{4}$$

$$\text{F1-score} = 2 \times (\text{Recall} \times \text{Precision})/\text{Recall} + \text{Precision} \tag{5}$$

The number of true positive samples is TP, the number of true negative samples is TN, the number of false positive samples is FP, and the number of false negative samples is FN [32].

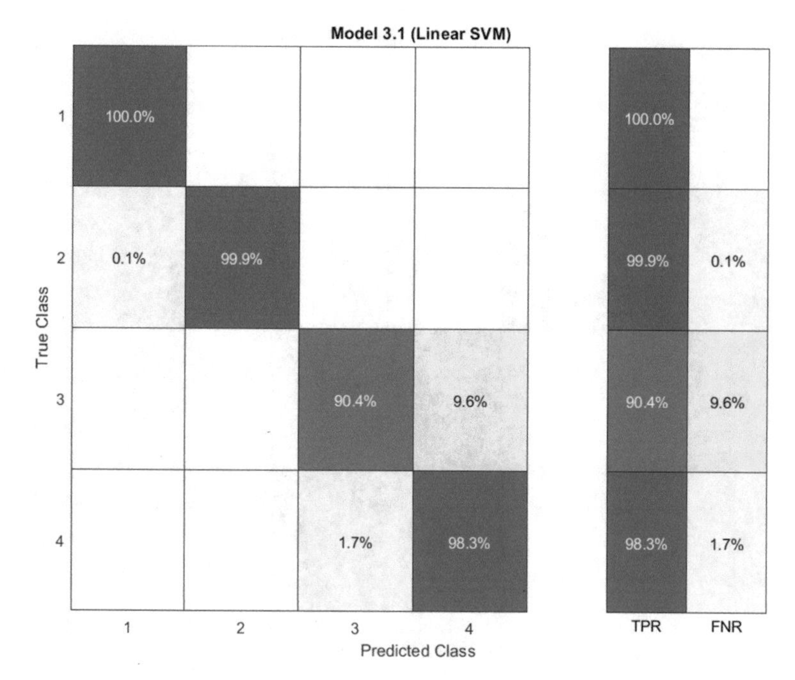

Fig. 2 Linear SVM confusion matrix

The datasets are splitted into training and testing sets for each helping component. We used the tenfold cross-validation strategy, the cross-validation selects the number of folds (or divisions) to partition the data set using the slider control, then the model execute.

5.1 Experiment I: Linear SVM

In the first experiment, linear SVM was used, the resulting output is depicted in Fig. 2. It is noticed that the class 3 is the most class that has miss classes while class 1 is the best one with o miss classes. The Fig. 2 also presents ratio of TP and FN. where class 1 gives the high ratio of TP.

5.2 Experiment II: Quadratic SVM

In the second experiment, quadratic SVM was used, the resulting output is depicted in Fig. 3. It is noticed that the class 3 is also the most class that has miss classes, but better than previous experiment, while class 1 is the best one with no miss classes.

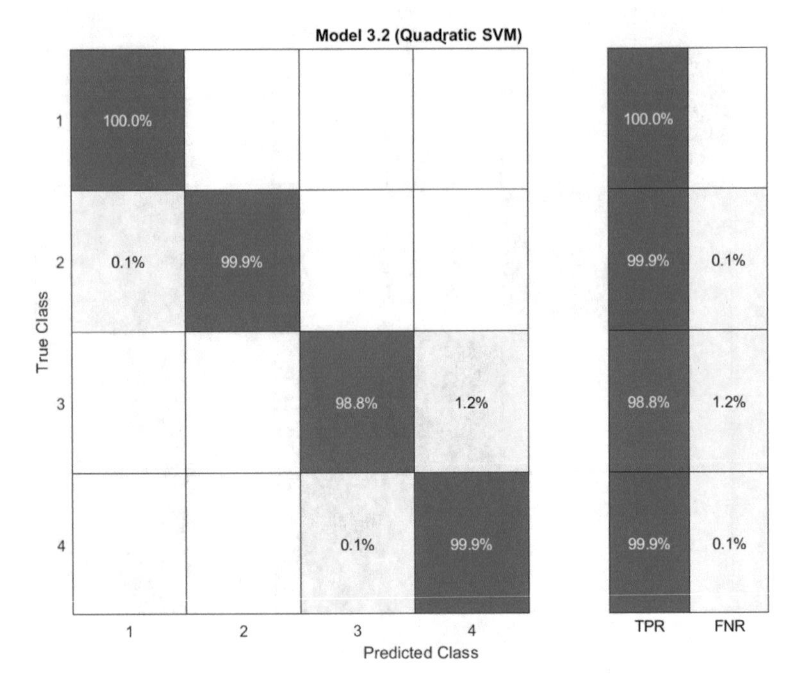

Fig. 3 Quadratic SVM confusion matrix

The Fig. 3 also presents ratio of TP and FN. where class 1 gives the high ratio of TP. The ratio of TP for class 3 is improved than previous experiment.

5.3 Experiment III: Cubic SVM

In the third experiment, cubic SVM was used, the resulting output is depicted in Fig. 4. It is noticed that the overall performance is improved where class 1 and class 2 is have no miss classes. The Fig. 4 also presents ratio of TP and FN. where class 1 gives the high ratio of TP. The ratio of TP for class 3 is improved than experiment I.

Table 3 shows the comparison of the experimental results for the three kernels. The linear SVM has an accuracy of 97.3, a prediction speed of 89,000 observations per second, and a training time of 15.968. The Quadratic SVM has an accuracy of 99.7, a prediction speed of 75,000 observations per second, and a training time of 11.529. The cubic SVM gives the best results, it has an accuracy of 99.9, a prediction speed of 76,000 observations per second, and a training time of 9.4245.

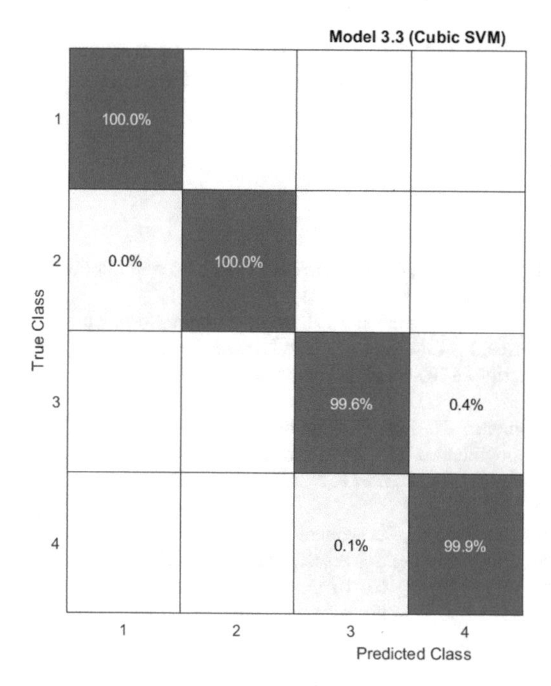

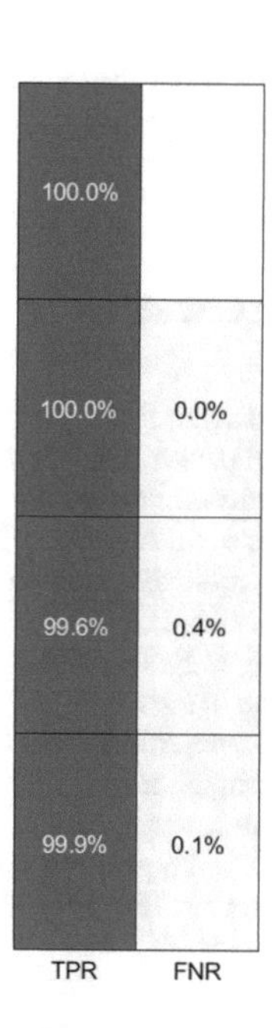

Fig. 4 Cubic SVM confusion matrix

Table 3 Comparasion of the experimetal results

	Accuracy	Prediction speed observation/(s)	Training time (s)
Linear SVM	97.4	89,000	15.968
Quadratic SVM	99.7	75,000	11.529
Cubic SVM	99.9	76,000	9.4245

6 Conclusion and Future Work

Metaverse allows user involvement. It is becoming a parallel realm where people work, play, and interacts. Metaverse uses mixed reality (MR) technology and ambient intelligence to connect digital and physical environments, allowing users to share resources. IoT lets the Metaverse fully integrate with the physical world. IoT creates an interoperable, seamless set of systems in the Metaverse to combine digital. As meat is essential for human health, yet it spoils easily. The proposed framework uses meta-veterinary to detect rotten meat in food security. IoT gas sensors and a machine-learning algorithm assess meat quality in the suggested method. The SVM method classifies meat grade using linear, cubic, and quadratic kernels. Performance was 97.4% linear, 99.4% quadratic, and 99.9% cubic. It is best to use cubic SVM because it is more accurate (99.9%) and takes less time to train. In further studies, we will investigate the potential educational applications of the Metaverse technology.

References

1. Banaeian Far, S., Imani Rad, A.: Applying digital twins in Metaverse: user interface, security and privacy challenges. J. Metaverse **2**(1), 8–16 (2022)
2. https://onpassive.com/blog/how-iot-and-metaverse-will-complement-each-other/
3. Darwish, A., Sarkar, M., Panerjee, S., Elhoseny, M., Ella Hassanien, A.: Exploring new vista of intelligent collaborative filtering: a restaurant recommendation paradigm. J. Comput. Sci. **27**, 168–182 (2018)
4. Eid, H.F., Darwish, A., Hassanien, A.E., Kim, T.: Intelligent hybrid anomaly network intrusion detection system. In: Communication and Networking. FGCN 2011. Communications in Computer and Information Science, vol. 265. FGCN 2011, Part I, CCIS 265, pp. 209–218. Springer, Berlin, Heidelberg (2011)
5. Anter, A.M., Moemen, Y.S., Darwish, A., Hassanien, A.E.: Multi-target QSAR modelling of chemo-genomic data analysis based on extreme learning machine. J. Knowl.-Based Syst., Elsevier, **188**, 104977 (2020). https://doi.org/10.1016/j.knosys.2019.104977
6. Kanter, T.G.: The Metaverse and extended reality with distributed IoT. IEEE Internet of Things Mag. (IoT) (2021)
7. Pereira, N., Rowe, A., Farb, M.W., Liang, I., Lu, E., Riebling, E.: Arena: the augmented reality edge networking architecture. In: 2021 IEEE International Symposium on Mixed and Augmented Reality (ISMAR), pp. 479–488. IEEE (2021)
8. Panea, B., Ripoll, G.: Quality and safety of meat products. Foods **9**(6), 803 (2020). https://doi.org/10.3390/foods9060803.PMID:32570790;PMCID:PMC7353557
9. Dionisio, J.D.N., Burns, W.G., Gilbert, R.: 3D virtual worlds and the Metaverse: current status and future possibilities. ACM Comput. Surv. **45**(3) (2013). https://doi.org/10.1145/2480741.2480751
10. Lee, L.-H., et al.: All one needs to know about Metaverse: a complete survey on technological singularity, virtual ecosystem, and research agenda, October 2021. http://arxiv.org/abs/2110.05352
11. Han, Y., et al.: A dynamic hierarchical framework for IoT-assisted Metaverse synchronization, March 2022. http://arxiv.org/abs/2203.03969
12. Gao, S.: Research on the innovation of the Internet of Things business model under the new scenario of Metaverse. In: Proceedings of the 2022 3rd International Conference on Internet and E-Business, pp. 44–49 (2022). https://doi.org/10.1145/3545897.3545904
13. Guan, J., Irizawa, J., Morris, A.: Extended reality and Internet of Things for hyper-connected Metaverse environments. In: 2022 IEEE Conference on Virtual Reality and 3D User Interfaces Abstracts and Workshops (VRW), pp. 163–168 (2022). https://doi.org/10.1109/VRW55335.2022.00043
14. Li, K., et al.: When Internet of Things meets Metaverse: convergence of physical and cyber worlds, August 2022. http://arxiv.org/abs/2208.13501
15. Seong-Soo, C.H.A.: Metaverse and the evolution of food and retail industry. Korean J. Food Health Converg. **8**(2), 1–6 (2022). https://doi.org/10.13106/kjfhc.2022.vol8.no2.1
16. Nam, H.W.: Metaverse environmental changes and technology trends. J. Korean Inst. Commun. Sci. **38**(9), 24–31 (2021)
17. Slovick, M.: The AR-VR age has begun in health care (2020, 11). https://www.cta.tech/Resources/i3-Magazine/i3-Issues/2020/November-December/The-AR-VR-Age-has-Begun-in-Health-Care

18. Metaverse Infrastructure: AI, IoT, digital twins, teleoperation and data management with support from 5G and beyond. https://www.researchandmarkets.com/reports/5566551/metaverse-infrastructure-ai-iot-digital-twins?gclid=CjwKCAiAjs2bBhACEiwALTBWZQCZeBnuHvf5ujOh77uZ2xpyyYKP9TjR11BHhonZ7fHItPJgwKkAqBoCUiYQAvD_BwE
19. Duan, H., Li, J., Fan, S., Lin, Z., Wu, X., Cai, W.: Metaverse for social good: a university campus prototype. In: Proceedings of the 29th ACM International Conference on Multimedia, pp. 153–161 (2021)
20. Shrouf, F., Ordieres, J., Miragliotta, G.: Smart factories in Industry 4.0: a review of the concept and of energy management approached in production based on the Internet of Things paradigm. In: Proceedings of the 2014 IEEE International Conference on Industrial Engineering and Engineering Management (IEEM), Selangor Darul Ehsan, Malaysia, 9–12 December 2014, pp. 697–701
21. Bandyopadhyay, D., Sen, J.: Internet of Things: applications and challenges in technology and standardization. Wirel. Pers. Commun. **58**, 49–69 (2011)
22. Atzori, L., Iera, A., Morabito, G.: The Internet of Things: a survey. Comput. Netw **54**, 2787–2805 (2010)
23. Motlagh, N.H., Bagaa, M., Taleb, T.: Energy and delay aware task assignment mechanism for UAV-based IoT platform. IEEE Internet Things J. **6**, 6523–6536 (2019)
24. Ning, H., Wang, H., Lin, Y., Wang, W., Dhelim, S., Farha, F., Ding, J., Daneshmand, M.: A survey on Metaverse: the state-of-the-art, technologies, applications, and challenges. arXiv preprint arXiv:2111.09673 (2021)
25. Sun, M., Xie, L., Liu, Y., Li, K., Jiang, B., Lu, Y., Yang, Y., Yu, H., Song, Y., Bai, C., Yang, D.: The Metaverse in current digital medicine. Clin. eHealth **5**, 52–57 (2022). https://doi.org/10.1016/j.ceh.2022.07.002
26. https://www.blockchain-council.org/metaverse/how-will-iot-integrate-the-real-world-with-the-metaverse/
27. Zhao, Y., Jiang, J., Chen, Y., Liu, R., Yang, Y., Xue, X., Chen, S.: Metaverse: perspectives from graphics, interactions and visualization. Vis. Inform. **6**(1), 56–67 (2022). https://doi.org/10.1016/j.visinf.2022.03.002
28. Liu, Z., Xu, H.: Kernel parameter selection for support vector machine classification. J. Algor. Comput. Technol. **8**(2), 163–177 (2014). https://doi.org/10.1260/1748-3018.8.2.163
29. Rahman Wijaya, D., Sarno, R., Zulaika, E.: Electronic nose dataset for beef quality monitoring in uncontrolled ambient conditions. Data Brief **21**, 2414–2420 (2018)
30. You, X., Wang, C.-X., Huang, J., Gao, X., Zhang, Z., Wang, M., Huang, Y., Zhang, C., Jiang, Y., Wang, J., et al.: Towards 6G wireless communication networks: vision, enabling technologies, and new paradigm shifts. Science China Inf. Sci. **64**, 1–74 (2021)
31. Wang, Y., Zhao, J.: Mobile edge computing, Metaverse, 6G wireless communications, artificial intelligence, and blockchain: survey and their convergence. arXiv:2209.14147v1 (2022)
32. Abouelmagd, L.M.: E-nose-based optimized ensemble learning for meat quality classification. J. Syst. Manag. Sci. **12**(1), 308–322 (2022). https://doi.org/10.33168/JSMS.2022.0122

A Framework for Shopping Based on Digital Twinning in the Metaverse World

Eman Ahmed, Ashraf Darwish, and Aboul Ella Hassanien

Abstract This chapter investigates how the customer's classical shopping experience can be migrated to the metaverse world using Digital Twinning Technology. This is in addition to explaining the problems that customers face in the classical shopping real-world model and how it can be alleviated in the metaverse. The main objective is to get an immersive shopping experience. Illustrations of the commerce models existing until now are presented. Our model demonstrates how offline retailing can be converted into an online one without the use of E-commerce or live commerce. Instead, Digital Twinning is used to convert offline retailing into an online experience similar to the real-world one. The use of Digital Twinning in providing planning opportunities for future retails is also presented.

Keywords Metaverse · Digital twins · Industry · E-commerce · Retail · Seller · Customer · Shopping

1 Introduction

Metaverse is a compound word of meta and universe, and it is a 3D world in which reality and virtual coexist [1]. Recently, various metaverse platforms are being launched in the market. In the metaverse, the users are able to create their own content beyond the limits of physical space, and experience things online which has been available only on offline.

The marketing potential of a metaverse lies in the ability to draw consumers through three-dimensional technology. Unlike e-commerce websites, metaverse and other technologies such as Augmented Reality (AR) can simulate real-world retail

E. Ahmed (✉) · A. E. Hassanien
Faculty of Computers and Artificial Intelligence, Cairo University, Cairo, Egypt
e-mail: e.ahmed@fci-cu.edu.eg

A. Darwish
Faculty of Science, Helwan University, Cairo, Egypt

E. Ahmed · A. Darwish · A. E. Hassanien
Scientific Research Group in Egypt (SRGE), Giza, Egypt

A. E. Hassanien et al. (eds.), *The Future of Metaverse in the Virtual Era and Physical World*, Studies in Big Data 123, https://doi.org/10.1007/978-3-031-29132-6_9

outlets [2]. This technology can show consumers navigation, interacting with products, and choosing among them. Instead of reading transaction logs, as done for online shopping, or even tracking how someone navigates through a site, marketers can now have a visual representation of consumer movements throughout the virtual store. Virtually, the experience of shopping can be seen as it is happening. Interaction between shoppers, employees and others can be seen and heard [3].

Many metaverse platforms have been launched such as The Sandbox, Decentraland, Voxels, and Somnium Space, and Meta's own Horizon Worlds. As reported by the BBC [4], many brands have been reserving locations in various metaverse worlds.

In the metaverse, one branch is to focus on making virtual world that mimic the real-world and another is to build a completely virtual world with virtual objects that don't exist in reality. For instance, special car designs may be found in the metaverse but not in reality. Similarly, Non-Fungible Tokens (NFTs) and smart collectibles can be obtained in the metaverse without having a real counterpart. But, imitating what already exist in reality represents a challenge especially when we try to link both worlds, the real one and the metaverse, together, such that a change in one affects the other. To do this Digital Twinning technology is necessary.

Digital twin technology enables us to build virtual models of physical objects in the real-world with a status similar to that of the physical counterpart using feedbacks from IoT sensors. It starts with building a virtual model using Modeling techniques such that it resembles its physical counterpart in all aspects such as the appearance, the texture, the size, the color, the geometric shape, and the functionality. Then, adding sensors, we can control and monitor the virtual object. In addition, experiments can be done on the virtual twin to be able to take decisions of what to do for the physical one. Digital twins have grasp attention in many fields including the retail management as in Maizi and Bendavid [5].

In this chapter, we would like to discuss the first branch of imitating the reality in the aspect of a shopping application. We are illustrating how to make an immersive classical shopping experience to the user using Digital Twin technology. In addition, we give hints how we can alleviate problems that exist in the real shopping experience. We also investigate the role of Digital Twinning in planning future shopping entities. As far as we know, this is the first paper to address the problem of how digital twins can be used in the metaverse shopping stores to provide better shopping experience to the users.

The chapter is organized as follows: Sect. 2 presents the basics and background. Section 3 demonstrates the related work. The classical shopping process is illustrated in Sect. 4 while Sect. 5 explains that in the metaverse. Section 6 demonstrates the proposed model using Digital Twinning. Section 7 elaborates the role of Digital Twinning in planning future retails. Section 8 discusses the challenges and potential solutions of such a system. The paper is concluded in Sect. 9.

2 Basics and Background

2.1 Digital Twin Overview

The Digital Twin is dependent on three primary components: data and information linkages, a physical product that is in a real-world environment that is properly monitored, and a virtual product existing in a virtual space. Firstly, sensors may be employed to maintain historical sites. Sensors perform monitoring in real-time and produce the huge volume of data. Following the transfer and storage of the data, it is then examined and linked to the virtual product, bringing to light information about how the physical space performs, how virtual space can be simulative, and how real-world decision-making can be done. All the different definitions of Digital Twin lead us to differentiate between 3 similar terms:

1. The digital model is the first term, referred to as a digital version of a pre-existing or future physical object. Here, there is not any automatic data exchange between the digital model and the physical object.
2. The second term is digital shadow, which is defined as an object's digital representation with only one direction of data flow: a change in the digital object is inferred by a change in the physical item's state.
3. Finally, the digital twin is the last term.

The point of contrast here is the full integration of the flow of data between the digital object and the existing physical object in the two directions Three primary components should be included in the digital twin: the physical object, its virtual model, and the data and information connections able to achieve a link between the virtual and physical objects.

Terms such as "digital shadow," "digital model," and "digital twin" may be employed while working with Digital Twin technology. Each of these terms has a distinct meaning that is not found elsewhere. Among them are the following:

1. Digital shadow is a digital replica of an object with data flowing from a physical object to the virtual representation in one direction.
2. A digital representation is a digital representation or replica of a real-world or planned thing. The physical thing and the model are unable to communicate with one another. This category will cover items such as product designs, building plans, and so on.
3. A digital twin exists when data flows between an existing physical thing and a digital object. Any change to the physical object impacts the digital object and vice versa.

To make a digital twin, Modeling and Simulation is used to create a virtual model that has the same features of the physical object. The features may include the appearance, the material, the size, the color, the texture, the functionality, and much more details that make a similar model. After that IoT sensors and actuators are used to get real-time feedback about the status of the physical object and reflect it to the

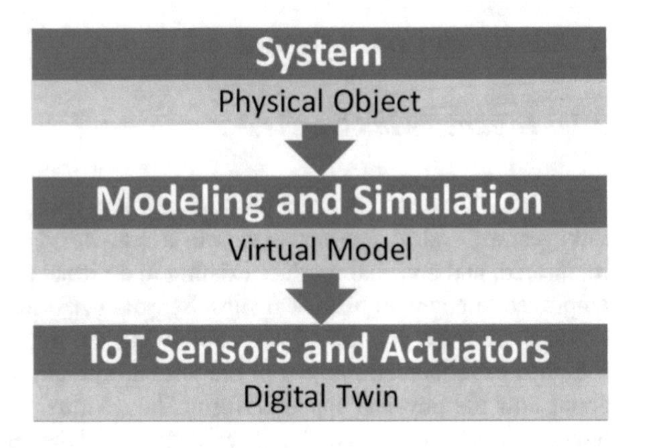

Fig. 1 Digital twin involved technologies

digital twin. Also, it can be used in the opposite direction, which is to control the physical counterpart from the digital twin. Figure 1 demonstrates how a digital twin is created.

2.2 Commerce Versus E-commerce Versus Live Commerce

Commerce is exchanging products between a seller a customer who pays for the purchase. In our world, sellers choose shops to present their products to the different customers who come to the shop. A shop usually has several partitions each responsible for presenting a certain category of products [6]. The products are usually put on shelves. The customers inspect the products then decide whether to buy them or not. A shop employee is available to help customers to find the products they want and guide them through the shop. In E-commerce, everything is virtual! The shop is a website consisting of webpages representing the partitions. There are menus from which the customer (the user in this case) can choose which partition to show. The products are images of the real products. They are shown inside tables instead of shelves. The assistants are virtual in the form of chatbots to which the customer can text and ask his questions and they reply. Recommendation systems can be used in E-commerce as well [7, 8]. E-commerce has benefits that it makes it easy for the customers to buy different products without going anywhere. It is easy to use. The purchase can be accomplished, and payment can be done both online using E-cards or in cash upon delivery. On the other hand, it has shortcomings that the user is not able to check products from all aspects, as only he could see images. The user is not able to check the material of the products for instance. Many faulty-size products can be delivered as well. The shopping experience in case of e-commerce is not immersive. Live commerce [9–11], on the other hand, is an emerging type of commerce incorporating live broadcast in which customers can see streaming videos of the shop and its products. It enables the user to check products as steaming videos instead

Table 1 Comparison between ordinary commerce, e-commerce and live commerce

Criteria	Commerce	E-commerce	Live commerce
Immersive experience	True	False	True
Products representation	Physical	Virtual images	Virtual streaming videos
Shop representation	Physical	Virtual website	Virtual
Payment	Cash or e-cards at cashier	Cash on delivery or e-cards	Cash or e-cards or crypto
Assistance	Human	Chatbot	Avatar

of images. It gives more immersive shopping experience. Table 1 summarizes the differences between ordinary commerce, e-commerce and live commerce.

3 Related Work

A digital twin prototype for in-store daily operations management is proposed and its impact on daily operations management performances was assessed on the fitting rooms' area. In [12], a semantic Digital Twin for Retail Logistics is proposed. The system is a semantically enhanced digital representation of a retail store. A scene graph demonstrates a realistic 3D model of the store with the shelf layout and contained products. DHL Trend Research Group [13] demonstrates how digital twins can play an important role in the logistics of several industries.

Gadalla et.al. [14] proposed a framework for the service quality of retails in the 3D internet or the metaverse. They inspected several dimensions that affect the quality including the customer service, the product, the store and the platform. In this work, a framework for retail service quality in the 3d internet is explained. The role of virtual environment has been investigated and detailed. The study demonstrated the importance of four dimensions for the retail service quality namely, customer service, product, store and platform. The features required for maintaining the service quality are mentioned for each dimension.

Regarding e-commerce in the metaverse, a business model for a new live commerce platform called MBUS is proposed in Moreira et al. [15] The users of the platform are the sellers who would like to present their products for selling and the customers who would buy the products.

4 Classical Shopping Experience

In this section, we explain the classical shopping in real-life and the problems that user face with it. The shopping process has two main players who are the seller and the customer. The seller initially buys a shop in a location and present his products in it so that users can see them and purchase what they need from them.

Most of the sellers choose their shops to be located inside shopping centers. This is because the shopping center collects many shops together, which make it easier to customers to go directly to one of them, check varieties of products of different shops and choose to buy whatever they want. In this paper, we discuss the shopping process inside shops of shopping centers.

The shopping process starts with a customer who enters a shopping center with an intention of buying a certain product of interest. The customer navigates the shopping center to reach a shop that sells the product types of interest. After that, the customer enters the shop and meet a customer service who listens to know what the customer needs, checks whether the product is available at the shop or not, then, provides assistance to get the product if it is available. The customer tries the product, check its price and make a decision of whether to buy it or not. In case of purchasing, the customer go to the cashier and pay for the product and take it or ask for a delivery. There are some problems with the classical approach such as:

1. The crowd:
    a. In case of a crowded shop, the user may miss seeing an available product which decreases the opportunity of a successful shopping.
2. Having large queues at fitting rooms or at the cashier:
    a. In case of having a large number of customers, large queues cause high latency in buying products which affect the shopping process badly.
3. Opening and closing times:
    a. In reality, everything has opening and closing times, even the employees have certain working times and the customer is obliged to visit the shop at these hours.

5 Shopping in the Metaverse

This section explains the process of shopping in the metaverse and how it can help us overcome the problems of that in the real-world.

1. A metaverse user will be able to visit any shopping center and navigate its various shops all over the world.
2. The user can try any product and make orders from any shop around the world.

3. The user will be focused on choosing the product of interest without having a burden of going through large distances to find the product.
4. Metaverse will enable us to alleviate shortcomings of shopping in the real-world, this includes:

    a. The crowd:
        a. The user can hide the other visitors and eliminate any crowd then look at all the products easily to choose among them. At the same time, the user can select to make some friends to accompany him/her and only show them in the shop.
    b. Having large queues at fitting rooms or at the cashier:
        a. This can be enhanced in the metaverse where the user is not obliged to wait in a queue to try a product or pay for it. Instead, he will be able to do the operation directly in no time.
    c. Opening and closing times:
        a. In the metaverse, these timings can be ignored, and the user can go shopping anytime of the day or even the holidays. This will enable more time for shopping. Only the delivery of products will be done in the opening hours of the shop in the real-world.

Figure 2 depicts the shopping scenario in the metaverse and its relation to the reality. As shown, the shopping process represents what happen in reality with additional features such as seeing a demo for a product and being able to try it. After purchasing and paying either using cryptocurrency or credit card, the role of the metaverse is over. After that the physical product is delivered to the customer in the real-world.

6 The Proposed Model of Shopping Using Digital Twinning in the Metaverse

Digital Twin technology plays an important role in providing the metaverse user an immersive shopping experience by mimicking the reality. Digital twinning will be used to build virtual models for products, shops, shopping centers and shop branches. Figure 3 demonstrates the digital twin types and their features. Next, we will illustrate the role of each digital twin to gain a better shopping experience.

6.1 Product Digital Twin

In the metaverse, we need to have a virtual model of the products that are found in the real-world. This would help in showing only products that are available at a shop at the time of shopping with the specs found in reality. In addition, any promotion that are happening in the real-world will be reflected simultaneously in the metaverse. Accordingly, each product should have a digital twin which includes features such

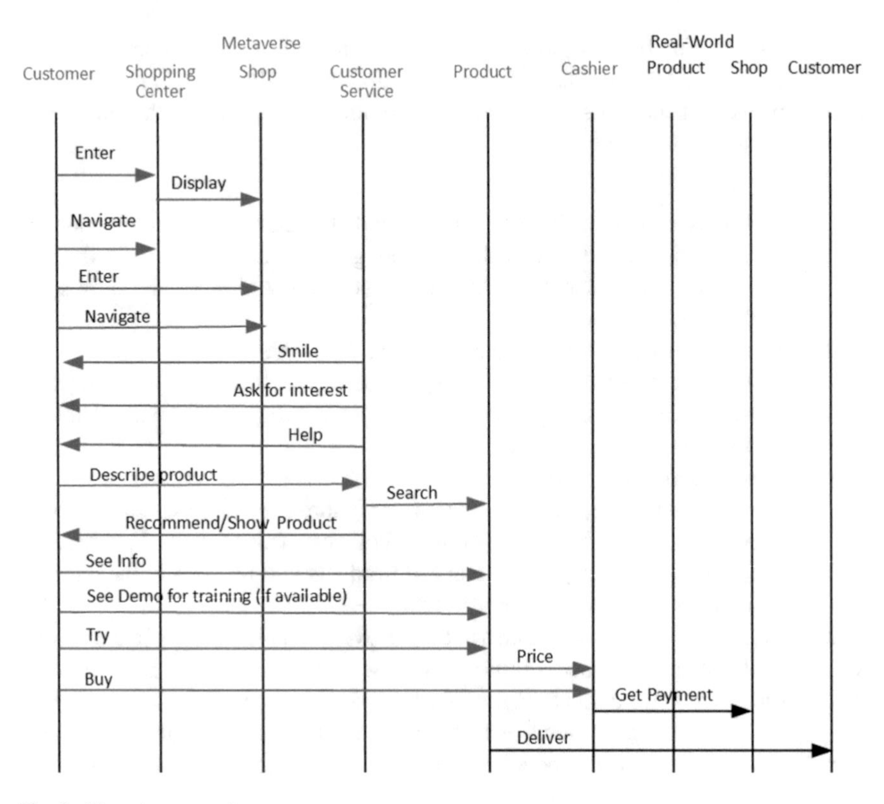

Fig. 2 Shopping scenario

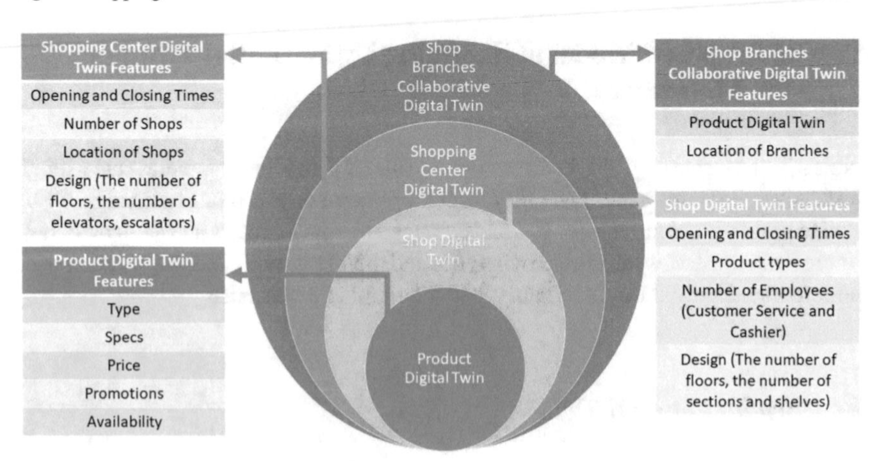

Fig. 3 Digital twin types

as its type, specs, price, whether it has a promotion or not, and its availability at the time of shopping. In case the shop sells a product and it is not currently available, the user shouldn't see it in the shop shelves, but, in case it is found in other branches, the user may be enabled to see it and shop branches collaborative digital twin will be used to handle the process as explained below.

6.2 Shop Digital Twin

For immersive experience, the shop design in the metaverse should resemble that in reality. Hence, information such as the internal design including number of floors, sections and shelves would be necessary. This is in addition to the product types sold by this shop in the real-world. Some features such as the opening and closing times, number of employees can be included. They can be used to mimic the reality in case of classical shopping experience. On the other hand, they can be used to test changing the timings and the number of employees to gain better experience in the real-world. Additionally, it can be a feature in the metaverse to ignore the opening and closing times, and to allocate a customer service and a cashier for each user in the metaverse regardless of what occurs in reality to provide a better shopping experience in the metaverse.

6.3 Shopping Center Digital Twin

The shopping center comprises many shops in different locations. It is useful to mimic the exact design of a shopping center. For instance, the shops in each floor and its location with respect to other shops. If shop A is next to shop B, it should be next to it in the metaverse. The navigation facilities such as elevators and escalators should be the same for an immersive experience.

6.4 Shop Branches Collaborative Digital Twin

In our previous work, we introduced the collaborative digital twins terminology to be defined as the integration of digital twins from different organizations to work together to achieve different objectives that benefit the society. In the case of shopping, collaborative digital twin will play an important role in demonstrating the shop branches. A product may not be available at one branch but found in another. In metaverse, the user may can have an option to see all the products available in all the branches, choose the product to buy, then collaborative digital twin of the shop branches will allocate the product to the nearest branch to the user location in reality and deliver the product efficiently from the nearest branch with reduced time, distance and fuel consumption.

7 The Role of Digital Twin Technology in Planning the Future Shopping Entities

On making decision on building a new shopping center in the real-world, it would be beneficial to first build a virtual model of its prototype with the proposed internal design, i.e. a digital twin, in the metaverse. The users will be able to navigate in it and provide feedbacks on its organization, design and even the theme. Updates in the design would be applied again to the digital twin and the process is repeated until the digital twin gets a design that provides an optimal shopping experience for the users. After that, building in reality will be accomplished.

Similarly, the same process can be used in designing new shops. Additionally, the promotions in shops can be tested in the metaverse prior to being applied in reality. This would give better information about the timing of the promotions and its type for increasing the selling rate.

8 Challenges and Opportunities

Creating digital twins at different levels is not an easy task. In addition, several challenges exist to create the mapping between offline stores and the metaverse digital twin. These are discussed in detail in this section. Also, opportunities for handling some of the challenges are explained.

8.1 Problems and Challenges

A. Digital Twinning problems include:

1. There are no standards for Digital Twin design, development, management, and operation, Digital Twin is difficult to copy, learn, and imitate. Various types of data transfer and interaction between different systems and devices are necessary during the process of developing and integrating Digital Twin models. As a result, developing data standards and communication interface protocols, as well as unifying data semantics and codes, is essential for constructing a comprehensive multidimensional Digital Twin platform. A significant barrier of Digital Twinning is the lack of homogeneity in existing communication interface protocols and data standards for many systems.

2. The Digital Twin model often consists of a mechanism or decision model that does not provide feedback or updates for management. Existing Digital Twin models are unable to accurately anticipate product life and manage dynamic operational processes after manufacture. Additionally, given the Digital Twin model's major objective, monitoring the product's health state and extending its life should be the key focus. Although there are significant gaps in the production and reuse stages, existing Digital Twin models are typically applied to the product's functioning period.
3. Digital Twin activities necessitate the processing of huge volumes of heterogeneous data, which represent difficult test of communication and storage capability.
4. It is difficult for different Digital Twin models to communicate with one another. Data's inconsistent semantic syntax causes redundant or missing knowledge resources, as well as impediments to basic knowledge base interoperability between levels.

B. Real-Time Database includes:
For creating the shop digital twin, we need to have real-time information about the products that are currently in the offline shop, their details such as their specifications as well as their positions among the shelves. When a product is moved from one position to another, this should be reflected too. When a product is sold, it should be removed from the shop database and the digital twin simultaneously. All these requires real-time database to reflect all the changes that are happening in the offline shop to its online digital twin. In case the product is not found at the current shop, the database must include information about its availability in other branches to choose the nearest one to get the product from. Once it is sold, the change must be done at the same time on the database of the collaborative digital twin. Accordingly, this is an essential requirement for the product, shop and the shop branches collaborative digital twins.
C. Real-Time Network Communication includes:
5G Networks must be used to enable real-time data exchange between the database and the digital twins. This is especially important for the product digital twin because it is the most dynamic. Also, it is an essential requirement for the collaborative digital twin to be able to send purchase orders from any branch and commit the selling process in no time.
D. Real-Time Computation:
It is necessary to have systems capable of delivering real-time computation for the digital twin to work effectively with the physical counterpart.

8.2 Potential Solutions

1. Standardization
A standardized framework that takes into account the interplay of platforms, software, interfaces, and technical rule coordination is necessary for the Digital Twin.

The Digital Twinning's general standardization effort is also in its early stages, and standard research content must be supplemented. Standardization content that may be addressed involves standards for fundamental aspects, concepts, technological implementation, testing, and assessment, as well as standards for collaboration across different systems. This is essential for Collaborative Digital Twin used for Shop branches to work effectively.

2. The use of Artificial Intelligence:

    a. Automation is necessary for the success of metaverse systems that mimic offline systems. It is really hard to rely on the human capabilities for making all the changes on the databases on real-time basis. Hence, AI can be an important choice to incorporate for automatic the process [11]. For instance, regarding the shop digital twin, AI can be used to categorize the products and help in the design of the shop regarding the shelves and the organization of the products such that similar products are put together. This can be done on the digital twin and reflect on the offline store for better shopping to the users in both metaverse and real-world. Similarly, this can be done for the shopping center digital twin, to organize shops selling similar products to be located in the same area, hence, make it easier for a customer to find the product s/he is looking for. A digital camera can be used to monitor the products, its status and register that on a database. AI can be used to analyze photos and determine when a product is removed from a shelf to indicate a possible purchase for this item in the digital twin. AI can also determine any defected product to be removed or make a suitable offer for it. Regarding the collaborative digital twin, AI can be incorporated in choosing the nearest branch and the best route it can take to reach the customer as soon as possible.
    b. Real-time computing is an issue for Digital Twining that must be taken into consideration and AI greatly aids in this area. It may effectively lower the difficulty of model creation and increase efficiency when used with machine learning.

3. Digital Twin models Migration
   The reuse of distinct models from various domains or scenarios inside the same domain is referred to as model migration. Model migration can minimize modeling complexity, accelerate model construction, and broaden model application in a variety of operating environments. After the modeling work is finished, model migration in many situations is considered for Digital Twin models with high accuracy.

9 Conclusion and Future Work

In this paper, we have explained how an immersive shopping experience can be provided to the customers of the metaverse. We shows how we can benefit from digital twinning technology to achieve our purpose. This is in addition to the important role for digital twinning in planning future shopping entities. In the future, we will investigate the usage of digital twinning technology in the metaverse in different application.

References

1. Suzuki, S., Kanematsu, H., Barry, D.M., Ogawa, N., Yajima, K., Nakahira, K.T., Shirai, T., Kawaguchi, M., Kobayashi, T., Yoshitake, M.: Virtual experiments in metaverse and their applications to collaborative projects: the framework and its significance. Procedia Comput. Sci. **176**, 2125–2132. Knowledge-Based and Intelligent Information & Engineering Systems: Proceedings of the 24th International Conference KES2020 (2020). https://www.sciencedirect.com/science/article/pii/S1877050920321529
2. Cagnina, M.R., Poian, M.: Beyond e-business models: the road to virtual worlds. Electron. Comm. Res. **9**, 49–75 (2009)
3. Swilley, E.: Moving virtual retail into reality: examining metaverse and augmented reality in the online shopping experience. In: Campbell, C., Ma, J.(J.) (eds.) Looking Forward, Looking Back: Drawing on the Past to Shape the Future of Marketing, pp. 675–677. Springer International Publishing, Cham (2016)
4. Murad, A., Smale, W.: The retailers setting up shop in the metaverse. https://www.bbc.com/news/business-61979150.html. Accessed 15 Dec 2022
5. Maizi, Y., Bendavid, Y.: Leveraging on the digital twin for improving retail store daily operations management. In: Proceedings of the 18th International Conference on Modelling and Applied Simulation (MAS), pp. 92–100 (2019)
6. Darwish, A., Sarkar, M., Panerjee, S., Elhoseny, M., Hassanien, A.E.: Exploring new vista of intelligent collaborative filtering: a restaurant recommendation paradigm. J. Comput. Sci. **27**, 168–182 (2018)
7. Moreira, Gabriel de Souza P., Rabhi, S., Ak, R., Kabir, Y., Oldridge, E.: Transformers with multi-modal features and post-fusion context for e-commerce session-based recommendation (2021). https://www.sciencedirect.com/science/article/pii/S1877050920321529
8. Parapar, J., Radlinski, F.: Towards unified metrics for accuracy and diversity for recommender systems. In: Proceedings of the 15th ACM Conference on Recommender Systems, RecSys 21, pp. 75–84, New York, NY, USA. Association for Computing Machinery (2021)
9. Merritt, K., Zhao, S.: The power of live stream commerce: a case study of how live stream commerce can be utilised in the traditional British retailing sector. J. Open Innov.: Technol. Market Complex. **8**(2) (2022). https://www.mdpi.com/2199-8531/8/2/71
10. Cai, J., Wohn, D.Y.: Live streaming commerce: uses and gratifications approach to understanding consumers' motivations. In: Hawaii International Conference on System Sciences (2019)
11. Eid, H.F., Darwish, A., Hassanien, A.E., Kim, T.: Intelligent hybrid anomaly network intrusion detection system. In: Communication and Networking in FGCN 2011. Communications in Computer and Information Science, vol. 265. FGCN 2011, Part I, CCIS 265, pp. 209–218. Springer, Berlin, Heidelberg (2011)
12. Kumpel, M., Mueller, C.A., Beetz, M.: Semantic digital twins for retail logistics. In: Freitag, M., Kotzab, H., Megow, N. (eds.) Dynamics in Logistics: Twenty-Five Years of Interdisciplinary Logistics Research in Bremen, Germany, pp. 129–153. Springer International Publishing, Cham (2021)
13. Dohrmann, K., Gesing, B., Ward, J.: Digital twins in logistics. DHL Customer Solutions and Innovation Represented by Matthias Heutger (2019). http://dhl.com/content/dam/dhl/global/core/documents/pdf/glo-core-digital-twins-in-logistics.pdf

14. Gadalla, E., Keeling, K., Abosag, I.: Metaverse-retail service quality: a future framework for retail service quality in the 3D internet. J. Mark. Manag. **29**(13–14), 1493–1517 (2013). https://www.sciencedirect.com/science/article/pii/S1877050920321529
15. Jeong, H., Yi, Y., Kim, D.: An innovative e-commerce platform incorporating Metaverse to live commerce. Int. J. Innov. Comput. Inform. Control **18**(1), 221–229 (2022). http://www.ijicic.org/ijicic-180117.pdf

Artificial Intelligence and the Metaverse: Present and Future Aspects

S. S. Thakur, Soma Bandyopadhyay, and Debabrata Datta

Abstract Artificial Intelligence (AI) and the metaverse are some of the most prominent technologies of the twenty-first century. Each can enhance people's lives in many ways, improve many industries, and increase the efficiency of many working processes. The metaverse is a compilation of virtual, augmented, and physical realities. Although "metaverse" terminology has been around for some time, it remains an emerging technology even as it becomes more of an everyday topic. Very soon, the metaverse will be a place where one can work, learn, shop, be entertained, and interact with others in ways that were never before possible. Without machine learning most metaverse experiences will not be possible which allows software applications to become more accurate at predicting outcomes without being explicitly programmed. AI and data science will be at the forefront of converging this combination of technology and will help revolutionize how people interact across the globe. It will lead to innovations, revenue streams, and deeper connections. AI and the metaverse can be applied in industries such as healthcare, gaming, management, marketing, education, and more. Generally, these technologies are looked at separately without considering their influence on each other and their potential for collaboration. In this paper, we talk about how artificial intelligence and the metaverse work together, how they are applied in various industries, and their potential. Before explaining the role of AI in the metaverse, its need to be explained the concept of the metaverse, where and how it is applied, and its potential.

Keywords Artificial intelligence · Metaverse · Web3 · Virtual reality · Machine learning

---

S. S. Thakur · S. Bandyopadhyay (✉) · D. Datta
MCKV Institute of Engineering, Howrah, West-Bengal, India
e-mail: somabanmuk@yahoo.co.in

S. S. Thakur
e-mail: subroto_thakur@yahoo.com

D. Datta
e-mail: debabrata.datta@heritageit.edu

D. Datta
Bhabha Atomic Research Centre, Mumbai 400085, India

A. E. Hassanien et al. (eds.), *The Future of Metaverse in the Virtual Era and Physical World*, Studies in Big Data 123, https://doi.org/10.1007/978-3-031-29132-6_10

1 Introduction

AI applications are now much more common than anyone might think. In a recent McKinsey survey, 50% of respondents said that their companies use AI for at least one business function. A Deloitte report found that 40% of enterprises have an organization-wide AI strategy in place [1]. In consumer-facing applications too, AI now plays a major role via facial recognition, natural language processing (NLP), faster computing, and all sorts of other under-the-hood processes.

It was only a matter of time until AI was applied to augmented and virtual reality to build smarter immersive worlds. AI has the potential to parse huge volumes of data at lightning speed to generate insights and drive action. Users can either leverage AI for decision-making (which is the case for most enterprise applications), or link AI with automation for low touch processes. The metaverse will use augmented and virtual reality (AR/VR) in combination with artificial intelligence and blockchain to create scalable and accurate virtual worlds.

The metaverse is expected to become the next big breakthrough in the Internet's evolution [2], with seemingly endless potential to transform how we live, transact, learn, and even benefit from government services. But making the metaverse a functional—albeit virtual—reality still requires significant advances in many different underlying technologies.

In some cases, these essential innovations are already underway. Meta, for instance, introduced an AI supercomputer in January 2022, claiming a range of uses from ultrafast gaming to instant, accurate and simultaneous translation of large amounts of text, images and videos [3]. The computer will also be key in developing next-generation AI models and become a foundation which future metaverse technologies can look to and build upon.

At the same time, consumer devices such as virtual reality (VR) headsets or smart glasses still fail to capture or transmit the full metaverse experience despite being available on the market today. New devices will need to be designed to provide a truly seamless experience for metaverse users. Another Meta invention, the haptic glove, is designed to enable users to touch and feel virtual objects in the metaverse [1, 3].

The Metaverse is one of the most significant rising technologies right now. In fact, the Metaverse is making such a stir in the virtual world that the social media behemoth Facebook even changed its official name to Meta to reflect the importance of the new universe. AI is likely to play a vital role in the inclusivity and accessibility of the Metaverse, making it more functional and user-friendly [4]. These technologies are shaping up to change the face of the internet and business in general.

This chapter is structured as follows. Section 2 describes the literature review. Section 3 presents the key promises of the Metaverse. Section 4 presents the challenges and future trends. Section 5 concludes and presents the future work of this chapter.

2 Literature Survey

The first record of the concept goes back to 1992 when Neal Stephenson brought the words meta and universe together in his science fiction novel Snowcrash [5]. In the book, the metaverse was described as a reality that people enter with the help of specific goggles and then interact with one another in this virtual space. In order to understand the metaverse, we first need to understand artificial intelligence (AI). AI is a process of programming computers to make decisions for themselves. This can be done in several ways, but the most common method is through machine learning [6, 7].

Machine learning (ML) is a subset of AI that allows computers to learn from data without being explicitly programmed [8–11]. This is done through a process of trial and error, as the computer tries to find patterns in the data it is given. Deep learning is a more advanced form of machine learning that uses artificial neural networks to learn from data. Neural networks are inspired by the way the human brain works, and they can learn tasks such as image recognition and natural language processing.

2.1 The Metaverse Versus VR

Many people mistake the metaverse and VR for synonyms; however, these concepts do not have much in common. VR, short for Virtual Reality, defines a simulated experience that people get through specific headsets. The situations in the VR world can be similar to the real world, such as meeting with friends, visiting the cinema, playing computer games, and more, or be completely different by immersing people in fictional and fantasy worlds. On the other hand, the term the metaverse is used to describe a hypothetical future representation of the internet that encompasses and is built around virtual reality, virtual worlds, and augmented reality. Currently, the metaverse isn't really there; however, it will potentially allow people to communicate, buy, sell, play, learn, work, and perform more activities through a centralized virtual experience [12].

To be more specific, in the real world, the metaverse describes a futuristic version of the internet that people will access through their VR headsets, augmented reality, and more common devices like mobile phones and personal computers or laptops. The difference between this concept and the existing web2 is enormous. Even the potential decentralized web3, where information is still accessed through websites and applications and communication happens via social media messengers and services like Zoom, Microsoft Teams, or Discord, is very different from the notion of the metaverse. The metaverse describes an immersive virtual world where users are represented by custom avatars and access information with the help of VR, AR, and other technologies [12].

Although the metaverse isn't here yet, some of its pieces exist in various forms. Some companies, such as Epic, for instance, started referring to some of their products as metaverses. For example, the battle royal game Fortnite is often considered a metaverse as the users can interact with the help of virtual avatars and the game displays various events by using VR technology.

2.2 Metaverse and Web3

One of the most recent vision about the Metaverse is the one published by Coinbase, which borrows the definition by venture capitalist and writer Matthew Ball: The future of the internet: A massively-scaled, persistent, interactive, and interoperable real-time platform comprised of interconnected virtual worlds where people can socialize, work, transact, play, and create [13].

The Metaverse is the distant evolution of Web3. In its most complete form, it will be a series of decentralized, interconnected virtual worlds with a fully functioning economy where people can do just about anything they can do in the physical world. A lot has been said about the internet and its future, but as Coinbase points out in its article, it is important to clearly separate the concept of the Metaverse from the concept of Web3 [14]. According to Chris, Web3 is about providing advanced digital services but, instead of these services being controlled by big technology corporations like in the Web2, they will be created and governed by the community, returning to the ethos of Web1, where the value of the internet was generated by the users at the edge of network, but mainly in a write mode. To be considered compliant with the Coinbase definition of a Metaverse, a platform should include the following elements/ characteristics:

1. Virtual Worlds
2. Massive Scalability
3. Persistency
4. Always on & Synchronicity
5. Platform to build upon
6. Fully functioning economy
7. Openness and decentralization
8. Interoperability.

Another vision of the Metaverse is the article by Jon Radoff [15], which also introduces the concept of the value chain of the Metaverse as presented by Jon Radoff, under CC BY license, as shown in Fig. 1.

Jon defines the Metaverse as "the real-time, activity-based Internet." Also, he points out that the Web3 is what enables value-exchange between applications in this new Internet the Metaverse will be [15, 16].

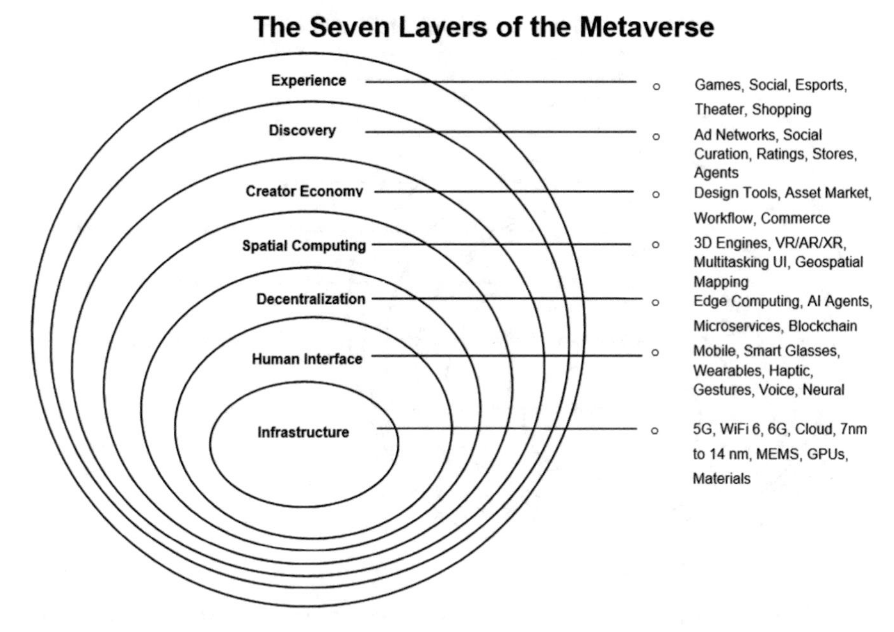

Fig. 1 The seven layers of the metaverse

2.3 AI in the Metaverse

There can be some areas in which AI can play a relevant role in the future of the metaverse, not only from a product perspective, but also considering how AI could make the Metaverse more inclusive. The layer schema provided by Jon [15] is depicted in Fig. 2.

3 The Key Promises of the Metaverse

The concept of the metaverse is based on some promises or functions this technology will be able to provide. These key promises include a decentralized world, identity verification, smart contracts, and ETPs. Let's define every goal to understand where AI can support and enhance the metaverse [5].

The first promise is a decentralized world that the metaverse represents. It is quite similar to the concept of web3, where users control their own data, digital assets, and identity, which are secure and tamperproof thanks to being stored on a distributed ledger. This goal will be the opposite of the existing web2 where few centralized tech companies hold the most power and will empower all users interacting with the internet. The next aim concerns identity verification. In this case, the metaverse implements blockchain technology to verify users' identities and ensure that only

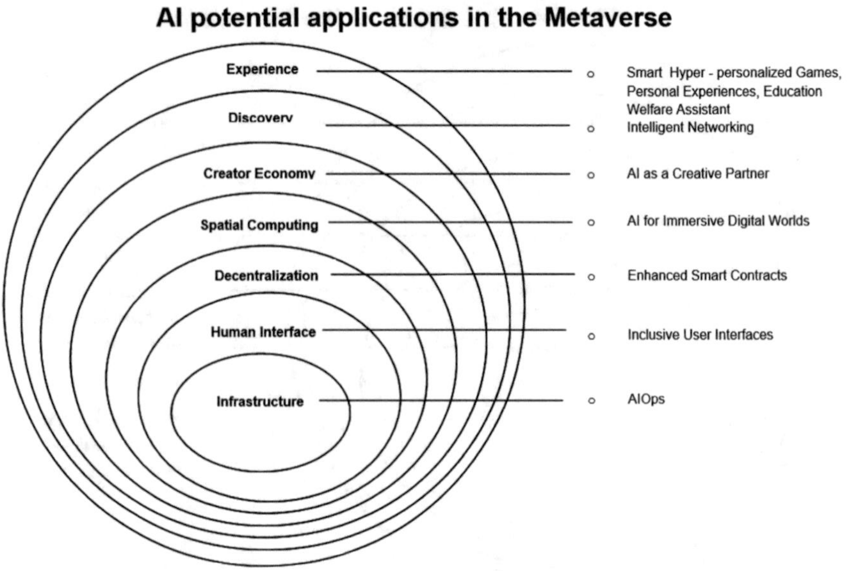

Fig. 2 AI potential applications in the Metaverse, based on "Building the Metaverse" by Jon Radoff, under CC BY license

users who passed the authorization can access data. Some of the leaders in the tech industry believe that the metaverse should implement self-sovereign identities (SSI) that are digital identities linked to verification and authorization data used in the real-world (e.g., biometrics) and managed in a decentralized way. One more promise the concept of the metaverse has is smart contracts. These contracts are used to automate and protect transactions. If we explain this technology more accurately, smart contracts are computer programs or transaction protocols that can automatically record, control, or execute actions based on the terms of a signed agreement or contract. Hence, the metaverse is expected to possess this function to ensure that various actions, such as trading, are executed according to the determined agreements. And last but not least, exchange-traded products (ETPs) are expected to be a part of the metaverse technology. This goal is quite simple to understand: the metaverse's original cryptocurrency will be used for buying-selling operations and rewarding AI applications for executing tasks.

Now, as we are well aware of the goals the metaverse has to achieve, we can proceed with the collaboration between AI and the metaverse. So, let's explain how artificial intelligence can bring humanity closer to creating it.

3.1 How AI and the Metaverse Collaborate

Chatbots are one of the greatest applications of AI that streamlines many business processes and allows users to solve their problems way faster. Also, this form of communication will be applied in the Metaverse. In addition to their current functions, such as customer service, marketing, sales, and more, chatbots can assist users in the Metaverse by providing them with instructions and information about various products and services, answering their questions, completing transactions on users' behalf, taking orders, etc. For instance, if a user cannot find a specific good, the chatbot could easily solve this problem by directing the customer to the needed location within the Metaverse [17, 18].

For more on chatbots, take a look at these AITJ articles: "Virtual Moron-Idiot!": Why Chatbots Fail and the #ChatbotRescue Mission Saving Them and Customer Service 2.0: The Rise of Chatbots and RPA. Also, one can learn more about the best chatbots here Best AI-Powered Chatbots [5].

Moreover, artificial intelligence can be utilized to create inclusive interfaces that will make the users' journeys convenient for everyone, including people with disabilities. Hence, AI can help make the Metaverse a user-friendly and easy-to-use platform. With the help of such technologies as Natural Language Processing (NLP), speech recognition, computer vision, translation, and augmented reality, the users will be able to interact with the Metaverse in their native language and through images and videos and enhance the user-metaverse interactions [19].

Digital avatars are one more way in which AI can collaborate with the Metaverse. For instance, to provide users with realistic avatars to represent them, AI can help create environments, dialogue, and images by using NLP, virtual reality, and computer vision. In a nutshell, AI can interact with the Metaverse in various ways, including digital avatars, chatbots, interfaces, and more. However, as the Metaverse is not here yet, artificial intelligence might progress even more until the new version of the Internet is here, which will bring new collaboration opportunities for these technologies.

3.2 Metaverse Versus Multiverse

Both the metaverse and multiverse are virtual spaces where people can interact with digital objects and one another in 3D environments. However, differences exist between the two. A metaverse is a holistic common shared place where all digital actions can occur, and one can enjoy various activities with others. In contrast, the multiverse is a pool of multiple virtual worlds. Metaverse activities might include playing video games, buying virtual real estate, shopping for real and virtual products, attending lectures or concerts, conducting 3D virtual video conferences, etc. On the other hand, a multiverse platform is more limited in that users can only use and interact within a single digital space for one specific purpose that the venue was designed for, for example, to play a particular game or exercise [3].

As mentioned previously, the metaverse is not a new concept, but then, again, it is not fully realized at this point. The metaverse is still evolving and looking toward a future state with astounding possibilities, and some even call this “the new internet.” Instead, we currently have multiple multiverses, each having its own attributes and integrated functionalities at different development levels. For example, Facebook already has one of the most known social media platforms and is moving toward creating a platform based on virtual and augmented reality. LinkedIn would be another example. Many multiverse platforms will converge as we move forward, bringing us closer to a “new internet.” This junction is where AI and data science will be critical in taking us to a new stage in our journey [20].

3.3 The Role of AI and Data Science in Shaping the Metaverse

The most important thing to remember about AI and data science is their power to parse tremendous volumes of data in real-time to provide meaningful insights. In addition, if executed properly, accurate interpretations will drive immediate action, interactions, responses, and low-touch automated processing. It will also include, among other things, event correlation, anomaly detection, and causality determination, thus helping to integrate and connect various platforms. The metaverse virtual reality must reflect physical reality more or less—the better connection to physical reality, the more realistic experience for the user. AI is the only tool that can accomplish this, and it would be impossible to achieve at scale by human means.

There are endless uses for AI and data science in the realm of the metaverse. Here are but a few:

1. 3D images
2. Animation
3. Speech
    a. Natural Language Processing (NLP), including real-time sentiment analysis, emotional analysis, and text classification.
    b. Translation
4. Design and artwork
5. Blockchain
    a. Smart contracts
    b. Decentralized ledgers
    c. Various transactions, including cryptocurrency.

3.4 Business Applications of the Metaverse

AIOps. AI-powered operations (AIOps) is a field of AI that uses ML to help businesses manage their IT infrastructure. AIOps can be used to predict and prevent outages, identify problems in real-time, and optimize resources. The metaverse's platform must be extremely robust, always on, and scalable, which is very demanding. Hence, AIOps can help to manage their resources more effectively and keep the systems running smoothly.

For example, if an organization's metaverse usage is spiking, AIOps can help to identify the issue and address it before it becomes a major problem [1].

AI bots. AI-powered chatbots are becoming increasingly popular for businesses. AI bots in the metaverse can be used for a variety of tasks, including customer service, marketing, and sales. It can help the users in the metaverse by giving instructions, providing information about products and services, answering customer questions, taking orders, and even performing transactions on behalf of the user.

For example, a customer might ask a chatbot where they can find a product. The bot would then be able to direct the customer to the correct location in the metaverse.

Inclusive user interfaces. The metaverse platform must be easy to use for all its users. This means that it must have a wide range of inclusive user interfaces, including those that are accessible for people with disabilities. AI for accessibility can help to create user interfaces that are more accessible for all users making the metaverse platform more user-friendly. AI technologies can be used such as:

1. Speech recognition to transcribe user speech into text and natural language processing to understand user intent
2. Computer vision to allow users to interact with the metaverse through images and videos
3. Translation to allow users to interact with the metaverse in their native language
4. Augmented reality to allow users to overlay digital information on the physical world [19].

For example, a user with a visual impairment might be able to use computer vision to see the metaverse in greater detail or a user with a hearing impairment might be able to use speech recognition to communicate with others in the metaverse.

Digital avatars. In the metaverse, users will interact with each other through digital avatars. These avatars will be used to represent users in the metaverse, and they will need to be realistic and lifelike. AI can help to create realistic avatars by using:

1. Computer vision to generate realistic images of users
2. Natural language processing to create realistic dialogue for avatars
3. Virtual reality to create realistic environments for avatars.

For example, a user might be able to create an avatar that looks like them, or they might be able to choose from a range of different avatars. The avatars could also communicate with each other, creating realistic dialogue. This would help to create a more social experience for users in the metaverse.

Digital twins. A digital twin is a virtual representation of a physical object. It can be used to track the status and performance of the physical object or to predict its future state. In the metaverse, digital twins will be used to represent objects in the virtual world. AI can help to create realistic digital twins by using:

1. Computer vision to create accurate models of physical objects
2. Deep learning to create realistic textures for digital twins
3. Virtual reality to create realistic environments for digital twins.

For example, a company might create a digital twin of their factory in the metaverse. This digital twin would be able to track the performance of the factory and predict its future issues. It would also be able to help with problems in the physical world by providing a virtual representation of the factory that could be used for troubleshooting.

New digital products and services: Businesses can leverage the metaverse space to create and sell new digital products and services. These could include:

1. Virtual reality experiences
2. Augmented reality experiences
3. Digital goods and services
4. Entertainment content.

AI can help businesses to create these new products and services by:

1. Analysing customer data to understand their needs and desires
2. Developing prototypes of new products and services
3. Creating realistic simulations of new products and services
4. Testing new products and services before they are released.

These can enhance customer experience, lower the cost of customer acquisition, or create new revenue streams, improve customer loyalty, and create a competitive advantage.

For example, a business might create a virtual reality experience that allows customers to explore their products such as a car or a house. This would give customers a better understanding of the product and could also increase sales.

Remote working and collaboration: The metaverse will enable people to work together from anywhere in the world. AI can help to facilitate remote working and collaboration by:

1. Enabling real-time translation of audio and text
2. Allowing users to share documents and files in the metaverse
3. Creating realistic simulations of meeting rooms and office spaces
4. Providing virtual meeting tools, such as whiteboards and video conferencing.

For example, a team of people might be able to work together on a project from different parts of the world by using the metaverse. They would be able to communicate with each other in real-time, share documents and files, and collaborate on projects. This would help to improve communication and collaboration between team members.

Enhanced smart contracts. The metaverse AI will come up with a platform that enables the exchange of digital assets and entitlements for the users. This will allow the protection of ownership and no interference from the big tech companies. Now, the bigger question is, will this be possible? The answer to this question can be stated as under; For instance, a huge brand called "Adidas" has of lately launched its first-ever NFT. However, the purchases were restricted to 2 people. What happened later on was that the sale was sold out in less than a second. The result was only one person was able to buy 330 commodities in a solo transaction. Now, this, in some way, is called the future of internet democratization.

It is clear that the power will be taken from the corporation houses, but the pertinent question is whether the power is in the hands of the people or not. The chances are a little dicey but in case it does, then a limited section of people will be able to access it. It is because there will be limited people who must have such high-level skills to retrieve benefits.

This type of layout will be laid out with the help of enhanced smart contracts, integrating AI. With regards to this, in the AIOps section, the infrastructure details corresponding to the transactions are helpful in finding out the "anti-democratic activities". Nonetheless, the blockchain transactions will not be halted after they have been completed.

Education and training. The metaverse will be a valuable tool for education and training. AI can help to make the metaverse more effective for education and training by:

1. Allowing users to learn in realistic virtual environments
2. Providing virtual teaching assistants
3. Offering educational courses.

For example, a person might be able to learn how to drive a car in a safe virtual environment. They would be able to take driving lessons, practice driving in different conditions, and even crash the car without any real-world consequences. This would help people to learn how to drive faster and safer.

3.5 AI Use Cases in the Metaverse

While VR worlds can technically exist without artificial intelligence, the combination of the two unlocks a whole new degree of verisimilitude. This could impact the following five use cases [1].

Accurate Avatar Creation. Users are at the centre of the metaverse and the accuracy of a user's avatar determines the quality of experience for them as well as other participants. An AI engine can analyse 2D user images or 3D scans to come up with a highly realistic simulated rendition. It can then plot a variety of facial expressions, emotions, hairstyles, features brought on by aging, etc. to make the avatar more dynamic. Companies like Ready Player Me are already using AI to help build avatars for the metaverse, and Meta is working on its own version of the technology[1, 20].

Digital Humans. Digital humans are 3D versions of chatbots that exist in the metaverse. They aren't replicas of another person, really—instead, they are more like AI-enabled non-playing characters (NPCs) in a video game that can react and respond to one's actions in a VR world. Digital humans are built entirely using AI tech, and are essential to the landscape of the metaverse. From NPCs in gameplay to automated assistants in VR workplaces, there are myriad applications, and companies like Unreal Engine and Soul Machines have already invested in this direction [20].

Multilingual Accessibility. One of the primary ways digital humans use AI is for language processing. Artificial intelligence can help break down natural languages like English, convert it into a machine-readable format, perform analysis, arrive at a response, convert the results back into English and send it to the user. This entire process takes a fraction of a second—just like a real conversation. The best part is that the results could be converted into any language, depending on the AI's training so that users from around the world can access the metaverse [1, 20].

VR World Expansion at Scale. Here's where AI really comes into its own. When an AI engine is fed with historical data, it learns from previous outputs and tries to come up with its own. The output of AI will get better each time, with new input, human feedback, as well as machine learning reinforcement. Eventually, the AI will be able to perform the task and provide output almost as well as human beings. Companies like NVIDIA are training AI to create entire virtual worlds. This breakthrough will be instrumental in driving scalability for the metaverse, as new worlds can be added without the intervention of humans.

Intuitive interfacing. Finally, AI can also assist in human–computer interactions (HCI). When one puts on a sophisticated, AI-enabled VR headset, its sensors will be able to read and predict one's electrical and muscular patterns to know exactly how he want to move inside the metaverse. AI can help recreate an authentic sense of touch in VR. It can also aid in voice-enabled navigation, so one can interact with virtual objects without having to use hand controllers.

3.6 Artificial Intelligence for the Extraordinary Digital World

The key elements that this platform provides to build virtual worlds and simulate authentic ones. They are amazing in the huge world-building and simulation of digital environments for examining self-governing robots to Artificial Intelligence enabled voice technologies. NVIDIA technologies are the best example of the creation of virtual spaces by the assimilation of AI with the metaverse to form social engagements. The gap between the virtual and the physical world has opened up a new domain for creators and marketers to explore and experiment with. The people are in huge demand for amazing experiences that are able to deliver important information along with inciting emotions. The two paramount characteristics called immersive media and voice are conveniently used to penetrate the ecosystem by satisfying the demand with fresh and meaningful experiences [21].

Digital twin campuses are all the rage, in light of Covid-19, anger of parents paying for a 'Zoom' education, and the advent of worlds built in virtual reality. Stanford has offered an anatomy class in VR and Arizona State has announced their intentions to do the same.

4 Challenges and Future Scope

Right now, the metaverse highly depends on VR (Virtual Reality), AR (Augmented Reality), and MR (Mixed Reality) technologies and devices. Since most of these are not lightweight, portable, or affordable, metaverse cannot have a wide-scale adoption. Apart from hardware accessibility, the challenge lies in having high-quality and high-performance models that can achieve the right retina display and pixel density for a realistic virtual immersion [22].

4.1 Challenges Around AI in the Metaverse

It is important to keep in mind that the metaverse is a new area of research and operation, and AI implementation could run into issues. For example, there could be questions around:

1. Ownership for AI-created content—Who holds the copyright to and can profit from the content and VR worlds created by AI?
2. Deepfakes and user transparency—How to ensure that users know they are interacting with AI and not other humans? How to prevent deep fakes and fraud?
3. Fair use of AI and ML—Can users legally apply AI/ML technologies to metaverse interactions? For instance, can they use AI code to win games?
4. Right to use data for AI model training—How can we ethically train AI for the metaverse? What are the consent mechanisms involved?
5. Accountability for AI bias—If a digital human or similar AI algorithm displays bias, what is the possible recourse?

Ultimately, without AI, it will be difficult to create an engaging, authentic, and scalable metaverse experience. That's why companies like Meta are working closely with think tanks and ethics groups to stem the risks of AI without curbing the technology's potential.

4.2 Challenges to Mainstreaming the Metaverse

Of course, a broad acceptance of the metaverse will depend on a positive user experience. Accessing the metaverse or any multiverse user requires a bulky headset, controllers, and high-speed connections, which can be somewhat costly. But every day, the movement to the metaverse requires more features and content that will create more and more demand, just as smartphones did just a little over a decade ago [23].

Another challenge is that initially, users may need to become a bit more tech-savvy. However, with emerging AI functionality, one can expect a world where it is simpler to join and experience the metaverse. At this time, there are countless uses for AI and data science to shape the metaverse and transform the online experience. As the metaverse evolves, AI and data science will have even more uses. In the future, there will undoubtedly be uses that are not conceivable today. IntellectData TM develops and implements software, software components, and software as a service (SaaS) for enterprise, desktop, web, mobile, cloud, IoT, wearables, and AR/VR environments [21].

5 Conclusion and Future Trends

The metaverse is a platform that has the potential to change the world. It is a platform where users can interact with each other in a virtual world, and it has the potential to revolutionize the way we work, learn, and play. AI can help the metaverse to live up to its promises by creating realistic avatars, developing new digital products and services, and facilitating remote working and collaboration. Businesses should look to the metaverse as a platform for innovation and creativity, and they should harness the power of AI to create new products and services that will improve customer experience and give them a competitive advantage. The metaverse is a virtual world that has the potential to change the way we live our lives. It can revolutionize the way we work, and AI can help to make this a reality. Experts predict that the metaverse will be based on seven essential technologies: 5G communication, extended reality, brain-computer interfaces, cloud computing, blockchain, digital twins, and artificial intelligence. Of these emerging technologies, AI may be the most crucial piece of the metaverse puzzle thanks to its potential to enable the metaverse to scale.

Deep learning-based software will likely power most interactions autonomously, with chatbots along with other types of natural language processing (NLP) technologies supporting all kinds of exchanges in this new extended reality space. AI will also enable machines behind the metaverse to not only understand user inputs, from text to images and even videos, but also to respond correctly regardless of the user's input language. However, this will require huge amounts of data and training such advanced NLP models is likely to take years. In metaverse development, AI is not only a necessary technology in the areas of computer vision and natural language processing, but also in VR and augmented reality (AR). For example, in AR technology AI is used in camera calibration, detection, tracking, camera pose estimation, immerse rendering, real-world object detection, virtual object detection, and 3D object reconstruction, helping to guarantee the variety and usability of AR applications. Eventually, most 3D images, animations, and speech in the metaverse will likely be generated by AI. Machine learning models could also be used to automate smart contracts, distributed ledgers and support other blockchain technologies to allow virtual transactions.

AI technology is also expected to help expand the metaverse by supporting object detection, improving rendering, and enabling cybersickness control and measurement. However, despite its promise and potential, the metaverse still has many challenges to overcome such as security risks and online abuse. Still, AI is likely to be among the technological tools that could be instrumental in overcoming those challenges.

To find out how current research is making the metaverse a reality, join ITU's fourteenth academic conference: Kaleidoscope 2022: Extended reality—How to boost quality of experience and interoperability, taking place in Accra, Ghana, from 7 to 9 December at the Ghana-India Kofi Annan Centre of Excellence in ICT. There have been significant advancements in machine learning, and the growth is expected to increase by leaps and bounds. The fusion of metaverse and artificial intelligence has all the possibilities to bring sci-fi to existence. It would be a place where people, apart from socializing, will be able to trade NFTs and other cryptocurrencies. There is only one thing that needs to be taken care of, which is too many platforms can make people confused. Much needed time and space should be provided in the upgradation process to make it more human-like.

References

1. Artificial Intelligence in the Metaverse: Bridging the Virtual and Real. https://www.xrtoday.com/virtual-reality/artificial-intelligence-in-the-metaverse-bridging-the-virtual-and-real/, *XR Today*, December (2021)
2. Cheng, S., Zhang, Y., Li, X., Yang, L., Yuan, X., Li, S.Z.: Roadmap toward the metaverse: an AI perspective. The Innovation **3**(5), 10029 (2022)
3. AI: The driving force behind the metaverse?, ITU News, www.itu.int/hub/2022/06/ai-driving-force-metaverse/, (2022)
4. Guo, Y., Yu, T., Wu, J., et al.: Artificial intelligence for metaverse: a framework. CAAI Artif. Intell. Res. **1**(1), 54–67 (2022)
5. Hryshkevich, H.: How AI and the metaverse work together. AI Time Journal, https://www.aitimejournal.com/how-ai-and-the-metaverse-worktogether#:~:text=Digital%20avatars%20are%20one%20more,virtual%20reality%2C%20and%20computer%20vision (2022)
6. Khan, L.U., Yaqoob, I., Salah, K., Hong, C.S., Niyato, D., Han, Z., Guizani, M.: Machine learning for metaverse-enabled wireless systems: vision, requirements, and challenges. arXiv preprint arXiv:2211.03703 (2022)
7. Anter, A.M., Moemen, Y.S., Darwish, A., Hassanien, A.E.: Multi-target QSAR modelling of chemo-genomic data analysis based on extreme learning machine. J. Knowl.-Based Syst., Elsevier, **188**, 104977 (2020). https://doi.org/10.1016/j.knosys.2019.104977

8. Salkuti, S.R.: A survey of big data and machine learning. Int. J. Electr. Comput. Eng. (2088–8708), **10**(1), (2020)
9. Abdelghafar, S., Darwish, A., Hassanien, A.E.: Intelligent health monitoring systems for space missions based on data mining techniques. In: Hassanien, A., Darwish, A., El-Askary, H. (eds.) Machine Learning and Data Mining in Aerospace Technology. Studies in Computational Intelligence, vol 836. Springer, Cham. https://doi.org/10.1007/978-3-030-20212-5_4
10. Sayed, G.I., Darwish, A., Hassanien, A.E.: Binary whale optimization algorithm and binary moth flame optimization with clustering algorithms for clinical breast cancer diagnoses. J Classif **37**, 66–96 (2020). https://doi.org/10.1007/s00357-018-9297-3
11. Darwish, A., Sarkar, M., Panerjee, S., Elhoseny, M., Hassanien, A.E.: Exploring new vista of intelligent collaborative filtering: a restaurant recommendation paradigm. J. Comput. Sci. **27**, 168–182 (2018)
12. Amos, Z.: How AI Is bringing the metaverse to life, https://www.unite.ai/how-ai-is-bringing-the-metaverse-to-life/, Augmented Reality (2022)
13. Pattam, A.: Intersection of metaverse and ai: how ai can help deliver metaverse promises? HCL Tech, April (2022)
14. Pereira, D.: How AI will shape the metaverse, towards data science. https://towardsdatascience.com/how-ai-will-shape-the-metaverse-4ea7ae20c99 (2021)
15. Radoff, J.: The metaverse value-chain, building the metaverse. https://medium.com/building-the-metaverse/the-metaverse-value-chain-afcf9e09e3a7 (2021)
16. Setiawan, K.D., Anthony, A., Meyliana, S.: The essential factor of metaverse for business based on 7 layers of metaverse—systematic literature review. In: International Conference on Information Management and Technology (ICIMTech) (2022), pp. 687–692. https://doi.org/10.1109/ICIMTech55957.2022.9915136
17. Pham, Q.V., Pham, X.Q., Nguyen, T.T., Han, Z., Kim, D.S.: Artificial intelligence for the metaverse: a survey. arXiv e-prints, arXiv-2202 (2022)
18. Ning, H., Wang, H., Lin, Y., Wang, W., Dhelim, S., Farha, F., Ding, J., Daneshmand, M.: A survey on metaverse: the state-of-the-art, technologies, applications, and challenges. arXiv preprint arXiv:2111.09673 (2021)
19. Lee, L.H., Braud, T., Zhou, P., Wang, L., Xu, D., Lin, Z., Kumar, A., Bermejo, C., Hui, P.: All one needs to know about metaverse: a complete survey on technological singularity, virtual ecosystem, and research agenda. arXiv preprint arXiv:2110.05352 (2021)
20. Moro Visconti, R.: From physical reality to the Internet and the metaverse: a multilayer network valuation. Available at SSRN (2022)
21. Role of Artificial Intelligence in Metaverse: Exploring the Saga of Metaverse with AI, Artificial Intelligence (2021). https://ai.plainenglish.io/role-of-artificial-intelligence-in-metaverse-22ae1cb679c3
22. Kohli, V., Tripathi, U., Chamola, V., Rout, B.K., Kanhere, S.S.: A review on virtual reality and augmented reality use-cases of brain computer interface based applications for smart cities. Microprocessors and Microsyst. **88** (2022)
23. Jagatheesaperumal, S.K., Ahmad, K., Al-Fuqaha, A., Qadir, J.: Advancing education through extended reality and internet of everything enabled metaverses: applications, challenges, and open issues. arXiv preprint arXiv:2207.01512(2022)

MultiModal Data Challenge in Metaverse Technology

Doaa Mohey El-Din, Aboul Ella Hassanein, and Ashraf Darwish

Abstract Metaverse becomes new trend around the world that is designed based on physical reality. Metaverse integrates several artificial intelligence technologies to simulate the real environments remotely and simultaneously that holds many characteristics to increase the virtual management systems. Metaverse is faced many obstacles in data science especially in data preparation, there is a different vision of future digital world known in Metaverse emerge the fusion data from multi modal. Recent motivations draw the vision of Metaverse future can inference in simulated virtual environments as the education, tourism, healthcare systems. So the importance of data science and data fusion from various sources/sensors to simulate the Metaverse systems has a great effect in the accuracy and performance of systems. Multi-modal data challenge in Metaverse technology is interpreted into powerful work. The various multimedia/multi modals input data such as Image, Text, audio, and video whether that related to spatial or temporal data. The more data I take, the more I know my son is a stronger, larger and integrated Metaverse model that are connected based on Internet-of-Things, digital twin, or blockchain technologies. This paper presents the importance of MultiModal data in construction Metaverse systems and recent motivations of multimodal in Metaverse. This paper presents a depth study about the Metaverse technology and the impact of data science in Metaverse and how the multimodal challenge effect on the accuracy of simulation of Metaverse with healthcare experiment. The experiment dataset includes 464 EEG, 464 MEG, and their derivates meta data of fMRI scan two types of excel sheets for brain analysis of neuro imagining. The fusion accuracy using Dempster-Shafer fusion technique results archives 57.91% accuracy of EEG, MEG, and fMRI. The fusion accuracy using concatenation fusion technique results achieves to 53.4% accuracy of EEG, MEG, and fMRI. The early fusion techniques improve the classification accuracy of multi-modal brain signals to improve the brain analysis with 3.5–6.5%. The early

D. M. El-Din (✉) · A. E. Hassanein · A. Darwish
Faculty of Computers and Artificial Intelligence, Cairo University, Cairo, Egypt
e-mail: d.mohey@alumni.fci-cu.edu.eg

A. Darwish
e-mail: ashraf.darwish.eg@ieee.org

Faculty of Science, Helwan University, Cairo, Egypt

A. E. Hassanien et al. (eds.), *The Future of Metaverse in the Virtual Era and Physical World*, Studies in Big Data 123, https://doi.org/10.1007/978-3-031-29132-6_11

fusion techniques improve the classification accuracy of multi-modal brain signals to improve the brain analysis with 7.5–9.5%. The Bayesian optimizer improves the accuracy results to 80 and 81%.

Keywords Metaverse · Digital twin · Virtual reality · Augmented reality · Avatar · Simulation

1 Introduction

The Metaverse technology can allow several users to manage and simulate various environments or games [1]; Where user cannot distinguish between fantasy and reality for example gamming playing simulation, Where the virtual reality glasses allow the quality of watching videos on different platforms. Real Metaverse application in middle east are illustrated designing an "avatar" or a digital object that will be you in this virtual world, using the "work room" performed by Facebook within the new technology [1]. Any Metaverse simulated system depends on the gathering between virtual reality, augmented reality, and recent mixed reality (MR) to improve the powerful experiences and useful interoperable data with respect to centralized or decentralized system [2]. Metaverse term is known as two parts, the first part is defined by "Meta", refers to beyond (non-existent), and the second part is defined by "verse" from universe, interpreting the world (beyond the world). The essential concept of Internet development is described of a virtual environment that does not exist on the ground.

This term dates back to the novelist Neil Stephenson, who coined the term "Metaverse" in 1992 in his novel "Snow Crash", in which he imagined characters Live virtual meet in 3D buildings and other virtual reality environments. The essential determination of "Metaverse" becomes the future trend of corporate Industry. According to the Journal of New York Times In 2022, Metaverse idea is applied on the future of digital technology for great international companies. Recently, there is a Metaverse revolution in investment. The Metaverse is growing, bringing many new opportunities for businesses and creators. The Metaverse market is expected growing exponentially to 13.1% every year [3].

Metaverse term interprets difficult to consist of the virtual reality. Metaverse includes three parts of simulated reality that are shown in, virtual, augmented, and mixed. While the features and properties of the Metaverse contrast from case to case, it is, within the least hardness terms, a shared virtual space that's intuitive, immersive, and hyper-realistic. It would moreover incorporate various contexts customized avatars and advanced resources. Although many motivations in the Metaverse technology around the world due to develop of the Internet based on socializing, working, and playing as simulated avatars [2]. The main key of Metaverse technology is shown in management remotely, keeping time, and elicits several context's relationships via interconnection of digital smart sensors. The parallel digital interaction and interpretation of data are applied to improve the accuracy of simulation of Metaverse.

The importance of usage of the Metaverse technology becomes very important especially in global international events wars and Covid-19 pandemic that have a great impact. For example, "Microsoft, Meta, Google, and Nvidia". In 2020, the market applies the revolution in the investment in Metaverse market to invest 478.7 billion dollars and is forecasted increased to 800 billion dollars by 2024 [4]. The predicted usage of Metaverse in 2026 reaches 25% of people will take an hour in the Metaverse and 30% of companies will provide services and creates simulations Metaverse products. The statistics of using Metaverse of United States for adults achieve 74%. The invested usage of Meta (Facebook) produces better than $10 billion. The market investment cap of Web 2.0 Metaverse organization is $14.8 trillion [3, 4]. In 2022, according to real survey of 5,521 Metaverse gamers, the largest group of users of the top 3 games (Minecraft, Fortnite, and Roblox) are under the age of 20. In 2040, 54% of these specialists said that they anticipate the Metaverse WILL be a much-more-refined and genuinely fully-immersive, well-functioning viewpoint of existence for a half billion or more individuals all inclusive [5]. In 2040, expected recent statistics of the usage of Metaverse technology around the world achieves to 46% to improve the truly fully-immersive, well-functioning with respect to daily functions and activities. The Metaverse isn't a single system but a collection of interconnected systems, each with its had rules and capabilities as shown in Fig. 1. It is made up of interconnected universes that are not bound by topographical boundaries, so, everybody can gotten the universe. The Metaverse has several processes to record the actions, tasks, mutations, real live functions in concurrent times. The Metaverse technology can support users in three dimensions involvement; that can support the virtual gatherings in gaming and works especially in "Corona" pandemic due to the hardness of real mobility. Metaverse becomes innovative alternative for virtual reality emergencies.

International Investment of Metaverse applications are happen [6]. Facebook is considered on the international organization of Metaverse revolution to support the computing construction applications. Metaverse has a new technology phase of interconnected virtual experiences which connected via Internet. The Metaverse has the

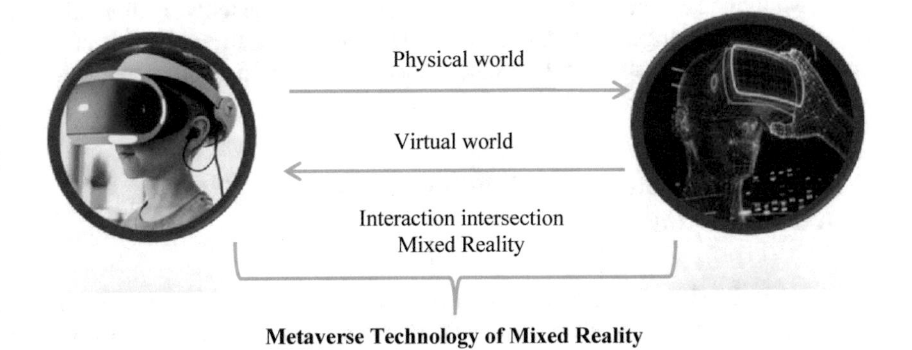

Fig. 1 Metaverse technology of mixed reality

potential to support conquer access to new innovative, social tasks and economic investment opportunities. And Europeans will be shaping it right from the start. After Facebook motivation improves the development technology companies as Microsoft, always shows to follow. Microsoft presents the big investment of several network technologies such as the security, spyware, video gaming, and hardware/mobile devices to perform a few. Business Metaverses simulations create free space in each room so that it can move around and fully integrate.

Metaverse motivations make a new revolution technology in MiddleEast [7]. Many countries apply the Metaverse technology in various applications. Healthcare can keep up and reconstruct of wellbeing by the treatment and avoidance of malady particularly by prepared and authorized experts [8]. Egypt goes forwards to apply the Metaverse technology on the medical tourism to do all the diverse activities, from work to sports and entertainment. "Emirates Health Services" launches the first "Metaverse" technology platform to serve customers during "Arab Health" [9, 10]. The new world of Metaverse in Saudi Arabia helps create real designs for a $500 billion city. Today, the Department of Technical and Digital Development of NEOM announced the XVRS platform, the "Dual Cognitive Digital Metaverse Platform," which is supposed to enable visitors to have a real-time presence in the city of NEOM, whether physical or virtual, such as avatars for an avatar or a hologram. The platform announced at the LEAP International Tech Conference in Riyadh that it is in the process of introducing a market for cryptocurrency and non-fungible tokens (NFTs). It's not all fun and recreations within the Metaverse. The tremendous world of virtual reality can moreover offer companies openings to discover arrangements online to apply to the genuine world. And these arrangements are what counseling firm PwC is trusting to make through its most current tech lab within the Middle East in Doha. This paper presents a depth study about the Metaverse technology and the impact of data science in Metaverse and how the multimodal challenge effect on the accuracy of simulation of Metaverse on Metaverse application.

The rest of the paper is organized as the following: Sect. 2, the Metaverse technology and the outlines of definitions, elements, features and conditions, Sect. 3, shows the data science of Metaverse analytics, Sect. 4, shows Metaverse applications, Sect. 5, Multi modal Data of Metaverse, Sect. 6, presents the benefits and limitations of Metaverse technology, Sect. 7, shows the Metaverse application experiment and results. Finally, Sect. 8 shows the conclude objectives the outlines of future works.

2 Metaverse Technology: Definitions, Elements, Features, and Conditions

Metaverse is considered the extension of the "real" virtual world that makes full simulation to reverse the real environments. Metaverse enables users to apply many interactions in the real time for improving simulated accuracy. Digital transformation is in the Web3 of adaptation to the digital world and its interactions becoming perceptible and real as in the physical world. The Metaverse targets the interoperability of open data standards among diverse simulated applications in different contexts [8]. A new digital behavior achieves to the freedom to display information from different resources in the same cooperative ecosystem in a virtual space. The main features of the Metaverse platform include its three-dimensional infrastructure, asset profiling system, and blockchain-based identity system (Fig. 2).

Virtual reality in digital advertising campaigns. In 2021, the global augmented and the investment size of virtual reality achieves 21.83 billion dollars. In 2030, according to the Institute in New York City, the excepted future prediction of Metaverse, large proportion of people will be in the Metaverse in some way on social media. In 2022, The Metaverse Group invests 3.2 million dollars. In 2027, NFTs is considered the key idea of the Metaverse system. Moreover, the Global News Wire, the global NFT market volume is expected to raise the investment *into 13.6 billion dollars* [9]. 55% of Respondents Expressed Concern About Privacy In Metaverse. The main dimension of the respondents according to Metaverse. The expected raising the market of Metaverse to 678.8 billion dollars by 2030. Lately, in this year 2022, Metaverse Market Size around the world will be increased to 47.48 billion dollars and 678. 8 billion dollars around the world by 2030 (Fig. 3) [11].

The Metaverse is picking up notoriety, and for great reason as well. The benefits of the Metaverse are simple to utilize of individuals to interconnect from anyplace over the globe amid the action and mimicked environment. Metaverse empowers to create more openings for both individual and commerce wanders. Innovation related to Web 3.0 such as blockchain and crypto monetary forms is portion of

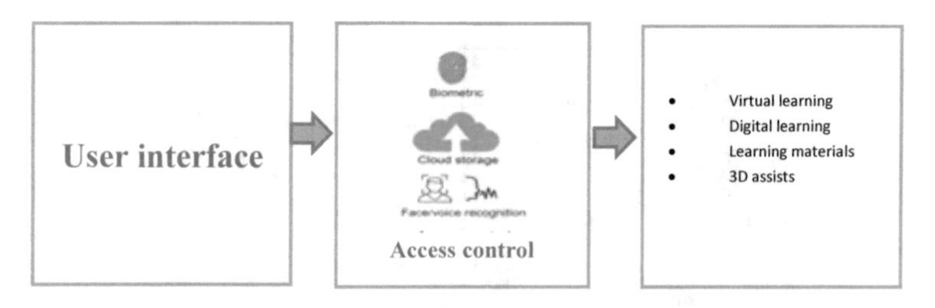

Fig. 2 Main architecture of Metaverse

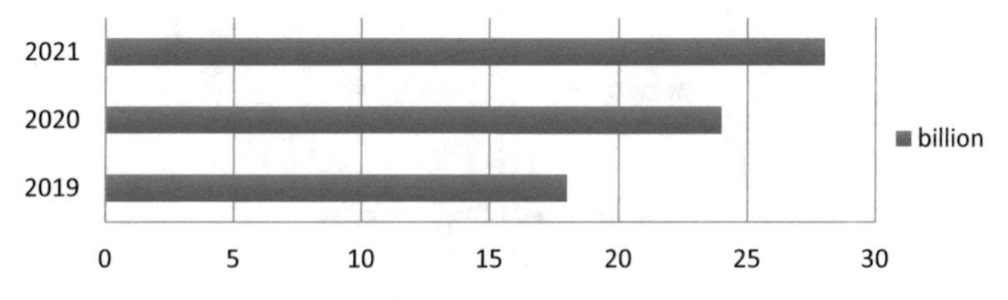

Fig. 3 The statistics of global augmented reality (AR), virtual reality (VR), and mixed reality (MR) market size from 2019 to 2021, according to statista [3]

the Metaverse encounter. Since Metaverse is built on blockchain, cryptocurrencies moreover utilize the same innovation. In this manner, it'll be the foremost well-known frame of installment within the Metaverse world. Utilizing cryptocurrency will be the only, most helpful, and slightest costly strategy to set out on a worldwide Metaverse shopping trip. You will utilize your cryptocurrency at any store, in any case of where it is found. You won't squander cash on universal communication or exchange costs on the off chance that you employ crypto, and most stores within the Metaverse acknowledge all major cryptocurrencies. Virtual reality and expanded reality are too related to the Metaverse.

There are the takings after six empowering innovations basic the Metaverse. With the development of miniaturized sensors, inserted innovation, and XR innovation, head-mounted shows or head protectors are anticipated to be the most terminal for entering the Metaverse. Computerized twin speaks to the advanced clone of objects and frameworks within the real world with tall judgment and awareness. Within the Metaverse, organizing innovations such as 6G, software-defined Network (SDN), and IoT engage the omnipresent organize get to and real-time enormous information transmission between genuine and virtual worlds, as well as between sub-Metaverses. Ubiquitous computing in specific context environment shows up anytime and all over for clients. Artificial intelligence has many innovations as the "brain" of Metaverse which engages personalized Metaverse administrations (e.g., striking and customized avatar creation), enormous Metaverse scene creation and rendering, multilingual back within the Metaverse by learning from authentic experiences (Fig. 4).

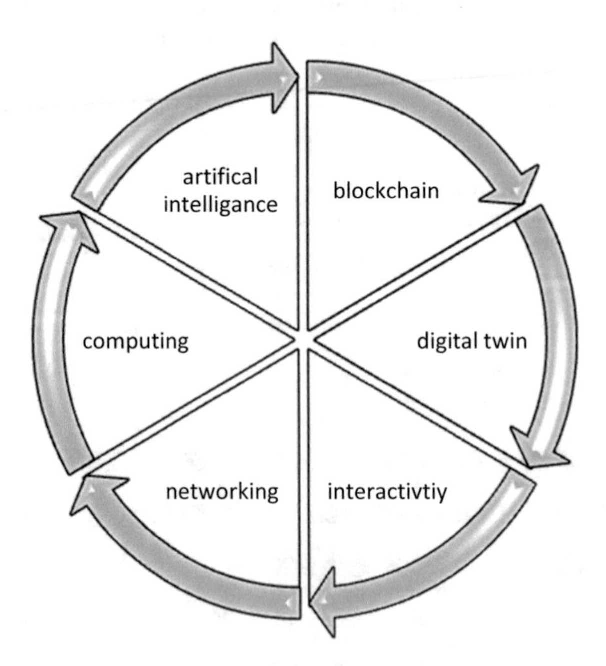

Fig. 4 The essential illustration of six used technologies in Metaverse

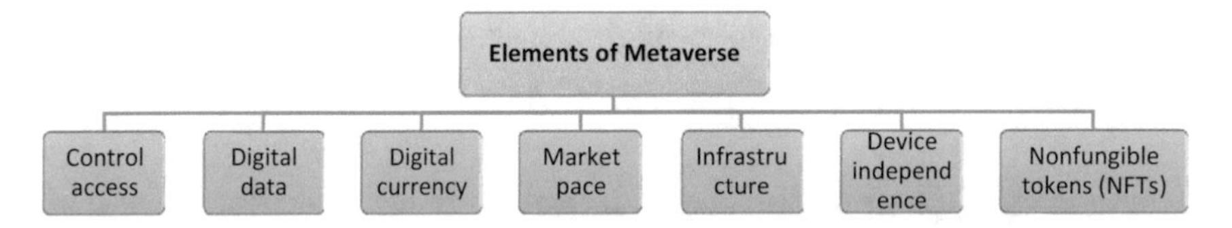

Fig. 5 Elements of Metaverse

Metaverse Elements include seven elements that are the digital currency, marketplace, infrastructure, device independence, Non-fungible Tokens (NFTS), and digital assets. Metaverse reproduces the digital experience equal to the physical experience, takes care of advantage to construct something unique and totally new through the digital twins. With this expands the opportunities for assets form a gateway for brands in the Metaverse of expanding brands to leverage their product (Fig. 5) [12].

By collecting real-time information and picking up experiences into how individuals move and carry on, able to make data-driven choices approximately how changing the basic foundation constructs the foundation that underpins future eras. Collaboration over the foundation industry and financial specialist is the key to building more intelligent, more feasible foundation. The early signs are that advanced twins themselves will move forward enormously, and to a degree, they are as of now doing permit for the reenactment or observing of options as the premise for speedier, more educated, and more dependable decision-making. When we collaborate, across boundaries, ready to do astounding things it's conceivable to form a way for better commerce choices that drive superior financial, social, and natural results. This vision of an interconnected computerized twinning of the built environment will have solid public-private computerized twins that can encourage the creation of show buildings or standard operations and administration protocol. To demonstrate scenarios that address supportability and versatility to make strides operational proficiency and create positive community results in open welfare, instruction, security, and financial advancement. This exact combination plans the virtual world into the genuine world in a cross-platform environment and applies the capacity to illustrate a assortment of practical situations. Giving generation proficiency for the producer to use virtual reenactment to create superior trade choices and create the most noteworthy return on venture. Several Successful components have contributed to this development:

(1) The sum of data being traded day by day premise has been expanding at the disturbing rate.
(2) The number of individuals getting to this data has been increasing.
(3) The number of gadgets that individuals utilize to get to this data has been expanding as well. The development of Virtual Universes has too been encouraged by innovative propels in computer equipment and computer program (e.g., quicker processors) as well as advancements in organizing innovations (e.g., Ethernet). The Metaverse isn't only vast, it is additionally energetic. It is always changing, and the changes are regularly unusual. It's not fair a put where individuals go to play games or socialize with companions. It is additionally a put

where individuals can go to memorize, conduct commerce, and indeed meet modern people.

3 Data Science of Metaverse Analytics

Data science and artificial intelligence improve the simulated development process, especially for the time-consuming and expensive processes like creating new integrations of sensors, smart devices, and assessing viable candidates [13]. The interpretation extracted motion big data for interconnection of Metaverse devices. Metaverse analytics refers to information gained from the computational analysis of data that includes various sets of data such as location data, interaction data, and sensor data.

Data science and artificial intelligence are driving forward massive environmental projects, including monitoring the climate, protecting wildlife, connecting people to nature, and much more. Revolutionizing climate estimates by building a computerized twin of our planet captures ceaseless and real-time information to supply profoundly exact gauges approximately extraordinary climate conditions, natural disasters, climate changes, and characteristic assets. Information science combined with machine learning technology underpins climate, agribusiness, biodiversity, and water ventures by making a difference monitor greenhouse gasses, clean up plastic and other flotsam and jetsam from the seas that hurt marine life, and protect natural life from numerous threats. The Information Science benefit supplier group can help businesses to contribute this vitality considering any of the continuous destructions they are finding in the current trade strategy and carrying out the alter in see of these illustrations so they can really plan for what is in store. The extended rule is likely progressing to have a huge effect on the courses of action which ought to be made. The foremost imperative thing to remember about AI and information science is their control (Fig. 6).

Machine learning and deep learning has several motivations to improve the classification accuracy for Multi-Modal Classification objects and Multi Modal Fusion objects [14]. Deep Learning Algorithms: The number of required connections in such a network rapidly grows to an unreasonable level as the input size increases the multimodal Fusion is pointed at utilizing the complementary data display in multimodal information by combining different modalities. One of the challenges of multimodal fusion is to expand inference to multimodal whereas keeping the demonstration and

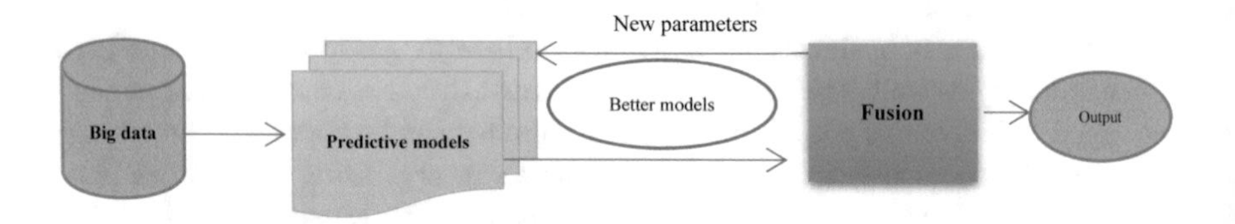

Fig. 6 Predictive analysis of Metaverse models

calculation complexity sensible The hardness of multi-data fusion process is shown in how to interpret a model for specific context based on many conditions and parameters. The Fusion of data in the smart Systems becomes a vital role in successful IoT operations: reporting analysis, devices connection, and remote management [15]. Each data source is different in target, organized architecture, input, context, and output.

4 Metaverse Applications

Metaverse holds many capacities to reproduce the real-world simulations utilizing revolutionary technologies like augmented reality, and virtual reality. Being the next iteration of internet and social media, Metaverse shows various business opportunities to enterprises worldwide.

For instance, Metaverse is going to take the advertising to another level using unique storytelling experiences in 3D technology (Fig. 7) [16].

- The interoperable concept in Metaverse project supports businesses to conduct and enter any event digitally.
- Brands can apply many interactions around the world via Metaverse technology to construct Metaverse frameworks.
- Metaverse also supports digital wallets that refers enterprises can regulate seamless transactions across its virtual ecosystem.

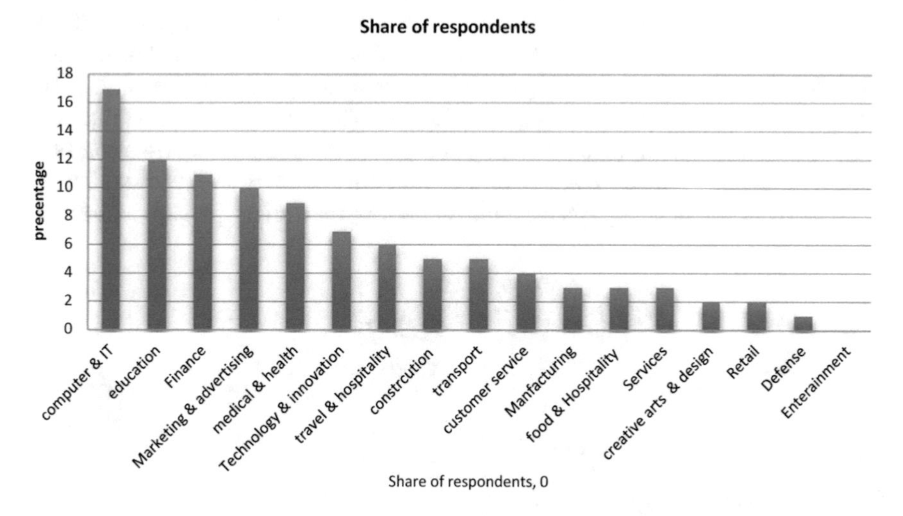

Fig. 7 Leading businesses sectors worldwide that have already invested in the Metaverse in 2022 [17]

In 2022, the annually survey report achieves to more than *17% of the global IT enterprises* have economic Metaverse simulated applications. In the interim other industry divisions such as Instruction, Back, Healthcare and Showcasing are rolling to work with Metaverse between 9 and 12%. Recent Metaverse applications are constructed for example Tenceni, Nvidia, Microsoft, Globant, Netease games, Alibaba cloud, Meta, Magic leap, Ueppelin. Recent motivations take care of the importance of interpreting fused data from extracted smart devices more usability and dynamically [17, 18] but they still face the challenge of high complexity.

5 Multi Modal Data Challenge of Metaverse

Metaverse applications apply on fourth multimedia with various characteristics and features [19] as Fig. 8. Recent motivations make a real simulation for five human senses as sight, sound, touch, taste, and smell. Advanced physics modeling applies various features and capabilities. Optimum physics-based modeling is simulated.

The fusion of Multimodal Data for real for sense improves the accuracy of simulated systems. The artificial intelligence models interpret the user/avatar behavior to simulate the sensed data to represent the physical users while waiting for users' responses. The data fusion integrity of the multi modal data inferences monitoring the sensors (hardware), data collection techniques, and artificial intelligence models (training and testing). Multimodal interaction can serve as a strategy to express the non-verbal cues, through employing various input sensors to capture our speeches, emotions, gestures, etc., and meanwhile delivering diversified (output) cues that stimulate our five senses, commonly recognized as visual, audio, and haptic feedback. Current contribution constructs a Metaverse simulator is based on Facial recognition Fingerprints Speaker (voice) recognition SMS (messaging service) Email [20]. The Avatar Turn-taking in a Virtual 'Metaverse' Planet. The Metaverse applies the association specks among multimodal enhanced interaction and Earthian-Martian non-real-time turn-takings. The key concept of usage the Metaverse is artificial intelligence of data driven embodied agents meet the Metaverse, the avatars, perhaps assisted by

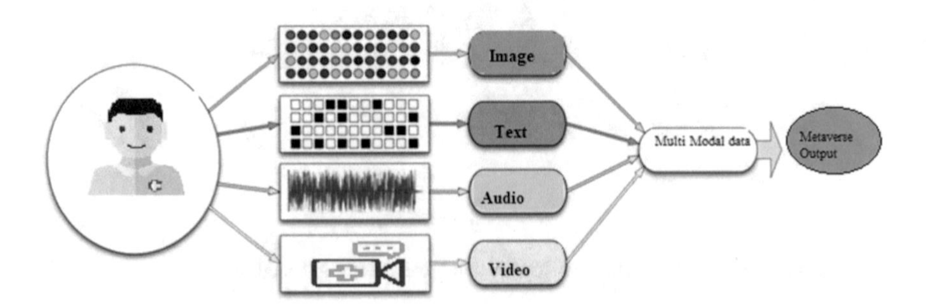

Fig. 8 Multi modal data for Metaverse simulation

artificial intelligence or intelligent organizations can interact with other avatars in for example an interplanetary virtual environment. In specific context, the captured data from the user's facial expression, physical body movements, and other internal states (e.g., heartbeat rates) can be transformed to a dataset expressing the user interaction trace mapping to conversation dialogues. The avatars, as a type of digital twins, can leverage the dataset to rebuild and simulate he management application of high-resolution user interaction via Metaverse technology based on multi modal inputs whether verbal textual input, images, or audio [21].

Algorithms can predict diverse units for each methodology: pixels or visual tokens for pictures, words for content and learned inventories of sounds for discourse. A collection of pixels is exceptionally distinctive from the sound waveform or a entry of content, as a result of this, calculation plan has been tied to a particular methodology. This implies that calculations are still working in an unexpected way in each modality. Data2vec shows the same self-supervised algorithm can work well in the diverse modalities [22]. Omnivore: A single model for images, videos, and 3D data [23]. FLAVA: A foundational model spanning dozens of multimodal tasks [24].

The essential effect of usage artificial intelligent provides the automated progress in Metaverse for operators and designers and delivers greater advantages over conventional techniques. Moreover, there is a lack of artificial intelligence interpretation of the streamline user operation and improve the immersive experience. Existing artificial intelligence models are quite deep and require the huge compute capabilities, making them unsuitable for resource constrained mobile devices [25]. As a result, building efficient artificial intelligent models is required. Blockchain interconnects the proof of work as its trust function. The blockchain can generate the interaction application in the Metaverse.

Current contributions take care of modality classification and modality fusion to be powerful and useful interpreted data from various modalities. Multi-Modality is also known by "Multi-Modal data" is interpreted into an inter-disciplinary approach that gets it communication and representation to be more than approximately dialect. It has been made over the past decade to deliberately address much-debated questions nearly changes in society, for event in association to cutting edge media and developments. The term of multi-modal refers to the combination of two or more modes in various combinations. The multi-modal applications include two phases,

(A) Modality Classification

Each modality has classification using text or image classifications. Machine learning and deep learning can classify objects from datasets to reach good accuracy.

(B) Multi-Modality Fusion

Multi-Modal includes fusion interpretation of modality types. There are three types of fusion, late fusion, early fusion, and recent motivation goes forwards to hybrid fusion. Early fusion has a lot features but not accuracy enough to train. Concatenation is one of early fusion techniques that uses for improving the accuracy with respect to most features. The late fusion has better accuracy results but that depends on classification learning step. Hybrid fusion achieves to the highest complexity but it has a redundant data. Dempster-Shafer is one of late fusion techniques that are a type of the Bayesian theory. It depends on the "probability mass function" and "confident interval" as shown in Eq. (1).

$$\left[\text{belief}, \ (A), \ \text{plausibility}_i(A)\right] \quad (1)$$

It is designed based on the plausibility confidence that means the upper level of the confidence interval, and it can count all the notices that do not revoke by the given proposition. The fusion rule is shown in Eq. (2).

$$\left(m_i \oplus m_j\right)(A) = \frac{1}{1-k}\sum\nolimits_{B\cap C=A=\emptyset} m1(B)m2(C) \quad (2)$$

Recent motivations go forwards to expect higher accuracy and computationally effective spatial and scene comprehension techniques to be utilized for the Metaverse in the near future. There is a great Transformation in the use of the smart environment and Metaverse in Health care [26].

6 Metaverse Benefits and Limitations

Metaverse benefits are classified into Fig. 9 as the following in businesses in more ways than one. It is interpreted based on the communication stable and better between smart sensors, easier, applies exercises, run the business, and educate like never before. The design of Metaverse application applies to specific context business and context requirements for several users [27].

Many Metaverse mechanisms of usage the virtual reality and artificial intelligence can simulate the physical lifes without losing the human connection. Recently, the usage of Metaverse is constructed based on crowd interviews or presentations for each specific context keys and community, that is has many benefits as the following,

1. **Communicate better**

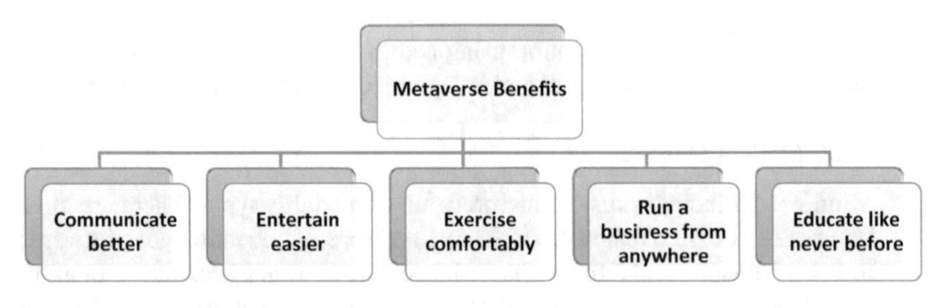

Fig. 9 Metaverse benefits

The interconnection communication of Metaverse can create 3D avatars and communicate and engage data. In the Metaverse, many tangible objects, like screens, will be holograms.

2. **Be easier**

   The Metaverse has great impact communities in many paths. In addition, to create the full vision of a virtual space in each context, and digital smart actuators or devices in the real world, enabling to interact with them. The Metaverse impact for simulating the entertainment context is great and variant issues in real-time. Moreover, Metaverse constructs to interconnect diverse devices into incorporated Augmented Reality experiments.

3. **Exercise comfortably**

   Many virtual reality technologies allow exercises in several properties. Metaverse enables users to make several experiences around the world.

4. **Run a business from anywhere**

   Virtual reality technology enables users to maintain a sense of presence and shared physical space. Moreover, mixed reality increases the productivity and enhances the global productivity and efficiency.

5. **Educate like never before**

   Metaverse depends on the virtual foundation to teach in a completely new and innovative path and enables users to learn from anywhere in the world.

6. **Examine the investment opportunities**

   Metaverse has a great impact of what is happen later.

Metaverse limitations are classified into two categories, Data and Network (Fig. 10). Data are collected from Virtual reality devices, sensitive information such as locations, shopping preferences and financial details, the hardness of track people to improve the degree than real world, recent motivations have illustrated that anonymous data can be identifiable. Risk of personality robbery, avatar duplication and abuse makes an issue for interoperability. Character confirmation build on square chain will be vital in this regard, because it is safer to cyber-attacks than a centralized framework [28, 29]. A decentralized recognizable proof arrange, empowering an account confirmation framework built on worldwide measures, to upgrade client certainty to use avatars over stages, might be one way to overcome this issue. Be that as it may, this would at the same time create bigger concentrations of information, making the accounts more helpless and potential harm within the occasion of a cyber-attack indeed bigger.

7 Experiment: A Multi Modal Data of Brain Signals on Face Actions for Healthcare Metaverse Applications

The interpreted dataset includes the healthcare Metaverse applications.

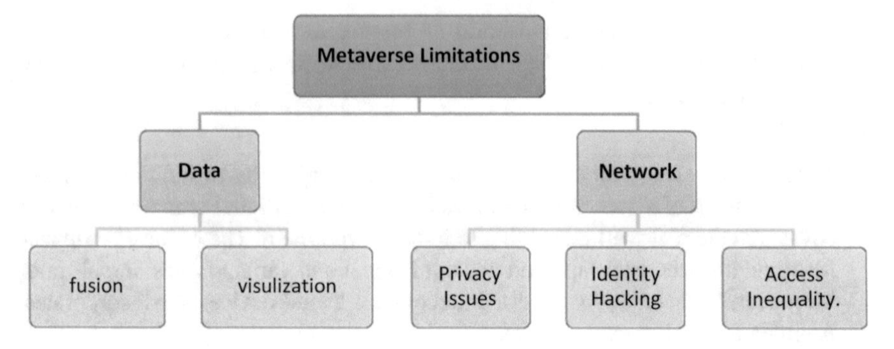

Fig. 10 Metaverse limitations

(a) *Dataset Description: is known as "A multi-subject, multi-modal human neuroimaging dataset"*

This dataset includes data acquired concerning to apply on multiple functional and structural neuroimaging modalities on the same nineteen healthy volunteers. The functional dataset contains three types of Neuro imaging that are Electroencephalography (EEG), Magnetoencephalography (MEG) and functional Magnetic Resonance Imaging (fMRI) data, recorded whereas the volunteers created numerous runs of hundreds of trials of a basic perceptual assignment on pictures of commonplace, new and mixed faces amid two visits to the research facility [30].

The structure of brain dataset contains T1-weighted MPRAGE, Multi-Echo FLASH and Diffusion-weighted MR sequences [31]. Though only from a small sample of volunteers, these information can be utilized to actualize strategies for intertwining numerous modalities from different runs on numerous members, with the objective of raising the spatial and transient determination over that of any one methodology alone. The combination estimations progress the basic network, and apply on the benchmark dataset of neuroimaging examination packages.

Electroencephalography (EEG) is an electrophysiological checking strategy to record electrical movement of the brain. EEG metrics voltage variances coming about from ionic streams inside the neurons of the brain. Diverse wavelengths from distinctive brain locales contain distinctive mental information. The analysis of MEG and EEG on 300 faces > On average across participants, 73% of popular faces were given a rating of 2–3, and 86% of non-famous faces were given a rating of 1. These questioning information may be utilized to assist refine recognition of each member with each confront, and are available on ask, but were not utilized within the current approval.

A useful MRI (fMRI) may be a more advanced sort of MRI that makes an energetic record of metabolic exercises over time. Functional-MRI innovation is centered only on the brain. MRIs show to users the interpretation of brain and other organs are arranged. fMRIs presents users the brain operations. The MRI data were gathered via Siemens 3T (Siemens, Erlangen, Germany). A standard 1 mm isotropic T1-weighted 'structural' image was acquired using an MPRAGE sequence (TR 2,250 ms, TE 2.98

ms, TI 900 ms, 190 Hz/pixel; flip angle 9°). The confront of the member within the T1 picture was subsequently physically expelled to assist keep up namelessness. The operational information were required utilizing an EPI arrangement of 33, 3 mm-thick pivotal cuts (TR 2,000 ms, TE 30 ms, flip point 78°). Cuts were required in an interleaves mold, with regard to the odd then even numbered cuts (where cut 1 was the foremost second rate cut) and a 25% separate dividing (raised where noteworthy to cover entire of cortex), coming about in a extend of voxel sizes of $3 \times 3 \times 3.75$ mm to $3 \times 3 \times 4.05$ mm over members. 210 volumes were required in each of 9 runs.

Magnetoencephalography, or MEG check, is an imaging strategy that distinguishes brain action and measures little attractive areas delivered within the brain. The filter is utilized to create an attractive source picture (MSI) to pinpoint the source of seizures as shown Fig. 11. The continuous MEG data interpreted using running that applies to MaxFilter 2.2 (Elekta Neuromag), that known as: (i) fitting a sphere to the digitized head points, excluding those on the nose, and utilizing the center of this sphere, collect the sensors' location, to identify a spherical harmonic basis set for Signal Space Separation (SSS) for delete the environmental noise27, (ii) programmed discovery of awful channels all through the run (extend over members and runs was from to 14, middle = 2), (iii) notch-filtering of the 50 Hz line-noise and its sounds, and of the HPI coil signals (from 293 to 321 Hz), (iv) compensating for development each 10 ms inside each run, (v) adjusting the information over runs to compare the head position at the begin of the fourth run (the run over members of development among runs was 0.2–7.9 mm, middle = 1.9 mm).

On the off chance that typically not the case spm_eeg_fuse makes it conceivable to combine two datasets with diverse channels into a single dataset given that the sets of channels don't cover and the datasets are indistinguishable within the other measurements (i.e. have the same testing rate and time pivot, the same number of trials and the same condition names within the same arrange). This function can be utilized to form a multimodal dataset too from independently recorded MEG and EEG which may be a substantial thing to do within the case that try with profoundly reproducible ERP/ERF.

(b) *Dataset size*:

The dataset size includes EEG, MEG and fMRI data that shows Figs. 12, 13, and 14. The dataset has four modalities, bi-modal images and bi-textual data input. Bi-modal images rely on the Neuro dataset has 464 Neuro imagining, and MEG 464 imaging. Bi-textual data has two excel files that have 8×49 table and 145×49 table.

(c) *Measurements*: the evaluation measures the classifier, three metrics of accuracy, sensitivity, and specificity were used as shown in Eq. (3) as follows [32]:

$$Accuracy = \frac{TP + TN}{TP + FN + TN + FP} \tag{3}$$

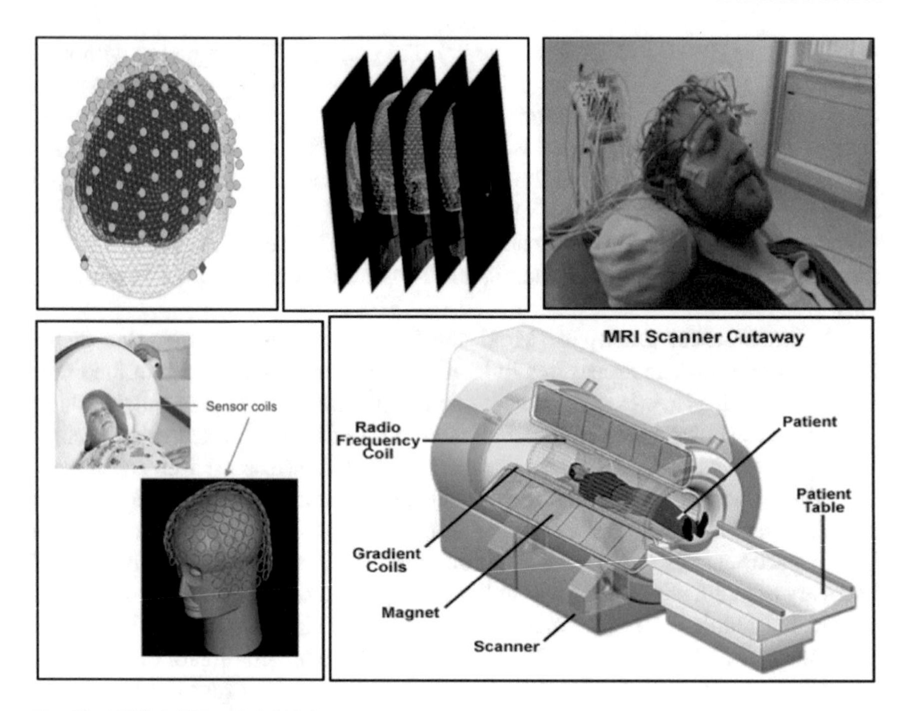

Fig. 11 EEG, MEG and fMRI data scans

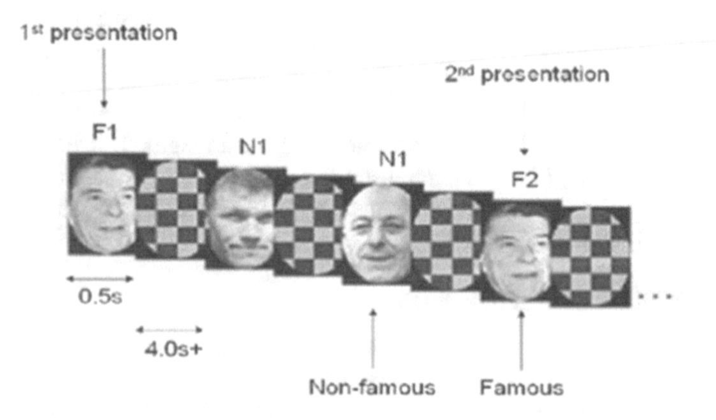

Fig. 12 Face repetition model

where *TP* refers to the right classifications of positive cases, *TN* interprets into the right classifications of negative cases, *FP* means the wrong classifications of negative cases into class positive, and *FN* refers to the incorrect classifications of positive cases into class negative.

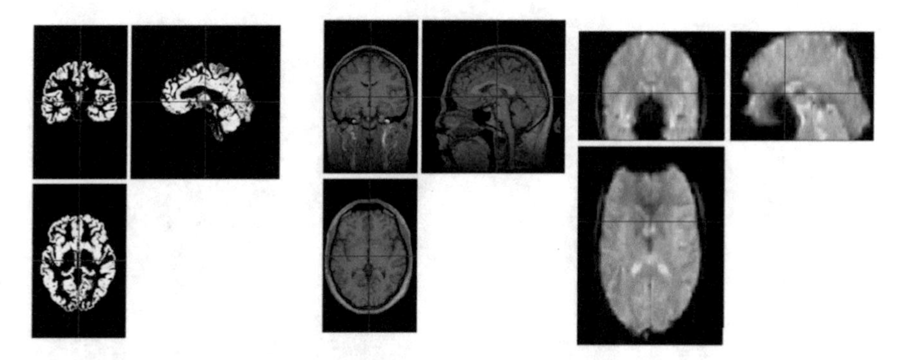

Fig. 13 Face repetition segmentation of structural image samples

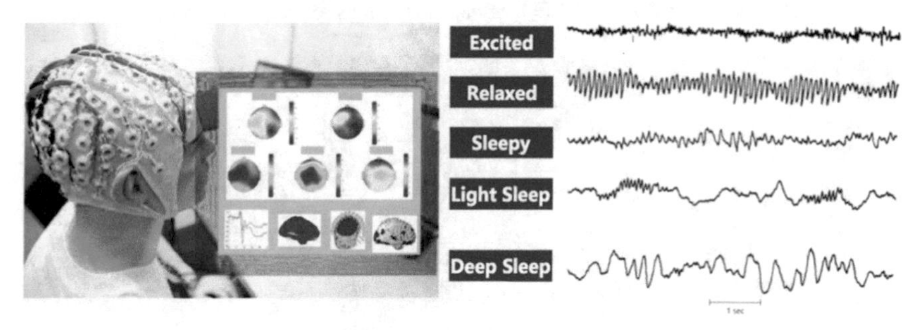

Fig. 14 Monitoring the whole brain enables to examine category specific brain activity

(d) *Results*

The experiment applies AlexNet classification using 22 layers on the neuroimaging dataset for three inputs as shown the sample in Fig. 15. Figure 16 presents a sample dataset of derivations data of MRI Neuro Imagining. A pre-trained Transfer learning of AlexNet is designed based on the convolutional neural network that constructs from 22 depth layers. It uses RelU activation function to improve the high speed impact of neural networks and uses Dropout in the fully-connected layers. In addition, the classification accuracy applies on a single modality as mention in Figs. 16, 17, and 18. The Fusion Classification accuracy of hybrid modalities using Concatenation fusion for multi-modal of Neuro imaging is as shown in Fig. 19. Fusion Classification accuracy of hybrid modalities using Dempster-Shafer fusion for multi-modal of Neuro-imaging is as shown in Fig. 20 and the sample of fused data in Fig. 21. Figure 22 presents the accuracy results of applying the Bayesian optimizer.

In both plots, at to begin with, the bend (classification exactness) climbs as the estimate of the ideally chosen highlight subset increments as shown in Figs. 15, 16, 17 and 18. It at that point remains around the extend of most extreme exactness

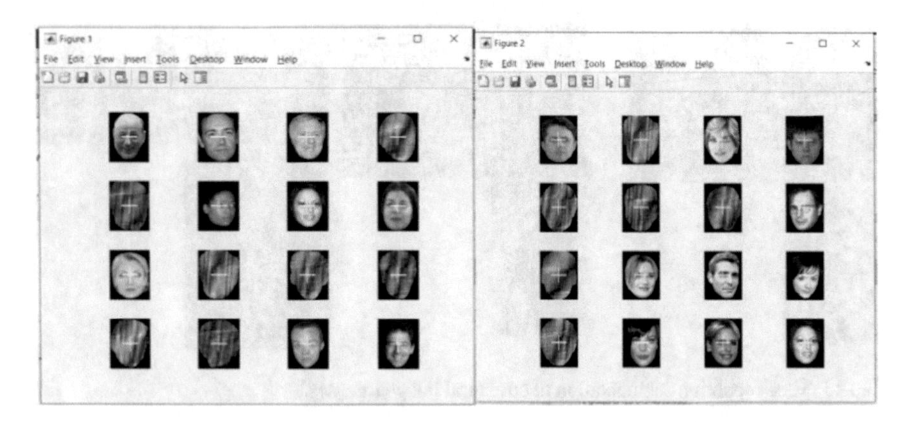

Fig. 15 A random sample dataset of EEG, MEG, MRI X-rays for brain imagining

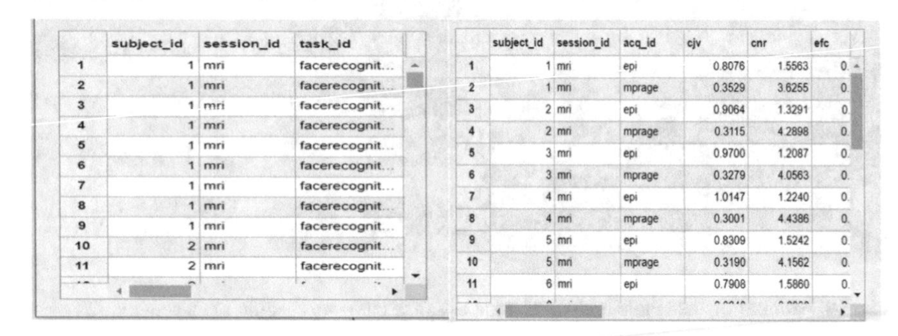

	subject_id	session_id	task_id
1	1	mri	facerecognit...
2	1	mri	facerecognit...
3	1	mri	facerecognit...
4	1	mri	facerecognit...
5	1	mri	facerecognit...
6	1	mri	facerecognit...
7	1	mri	facerecognit...
8	1	mri	facerecognit...
9	1	mri	facerecognit...
10	2	mri	facerecognit...
11	2	mri	facerecognit...

	subject_id	session_id	acq_id	cjv	cnr	efc
1	1	mri	epi	0.8076	1.5563	0.
2	1	mri	mprage	0.3529	3.6255	0.
3	2	mri	epi	0.9064	1.3291	0.
4	2	mri	mprage	0.3115	4.2898	0.
5	3	mri	epi	0.9700	1.2087	0.
6	3	mri	mprage	0.3279	4.0563	0.
7	4	mri	epi	1.0147	1.2240	0.
8	4	mri	mprage	0.3001	4.4386	0.
9	5	mri	epi	0.8309	1.5242	0.
10	5	mri	mprage	0.3190	4.1562	0.
11	6	mri	epi	0.7908	1.5860	0.

Fig. 16 A sample dataset of derivations data of MRI Neuro Imagining

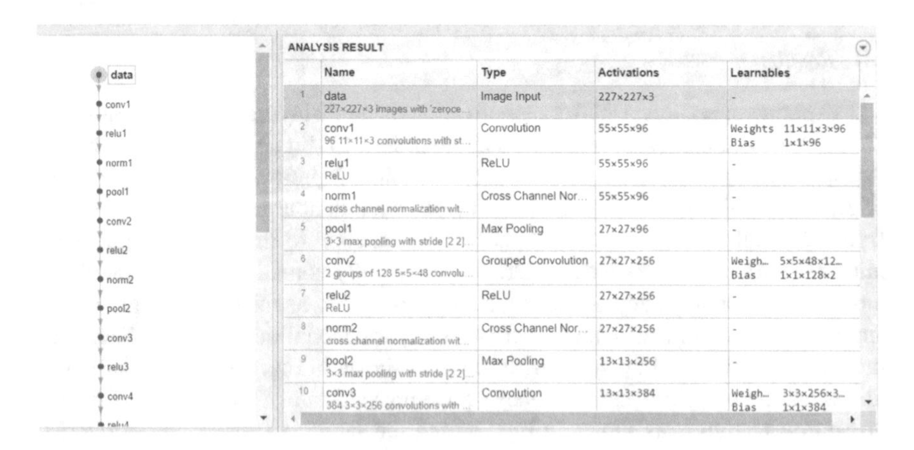

	Name	Type	Activations	Learnables
1	data 227×227×3 images with 'zeroce...	Image Input	227×227×3	-
2	conv1 96 11×11×3 convolutions with st...	Convolution	55×55×96	Weights 11×11×3×96 Bias 1×1×96
3	relu1 ReLU	ReLU	55×55×96	-
4	norm1 cross channel normalization wit...	Cross Channel Nor...	55×55×96	-
5	pool1 3×3 max pooling with stride [2 2]...	Max Pooling	27×27×96	-
6	conv2 2 groups of 128 5×5×48 convolu...	Grouped Convolution	27×27×256	Weigh... 5×5×48×12... Bias 1×1×128×2
7	relu2 ReLU	ReLU	27×27×256	-
8	norm2 cross channel normalization wit...	Cross Channel Nor...	27×27×256	-
9	pool2 3×3 max pooling with stride [2 2]...	Max Pooling	13×13×256	-
10	conv3 384 3×3×256 convolutions with ...	Convolution	13×13×384	Weigh... 3×3×256×3... Bias 1×1×384

Fig. 17 AlexNet neural network for Neuro imagining

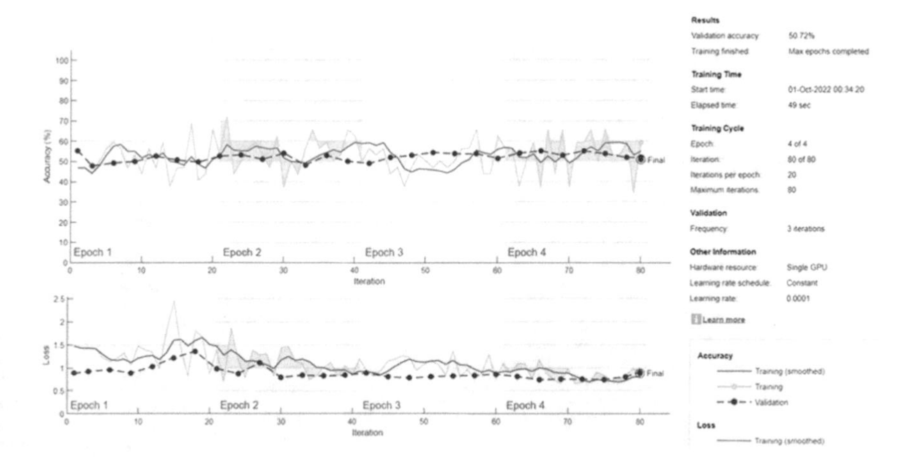

Fig. 18 Classification accuracy of single modalities for MEG Neuro imaging (50.72%)

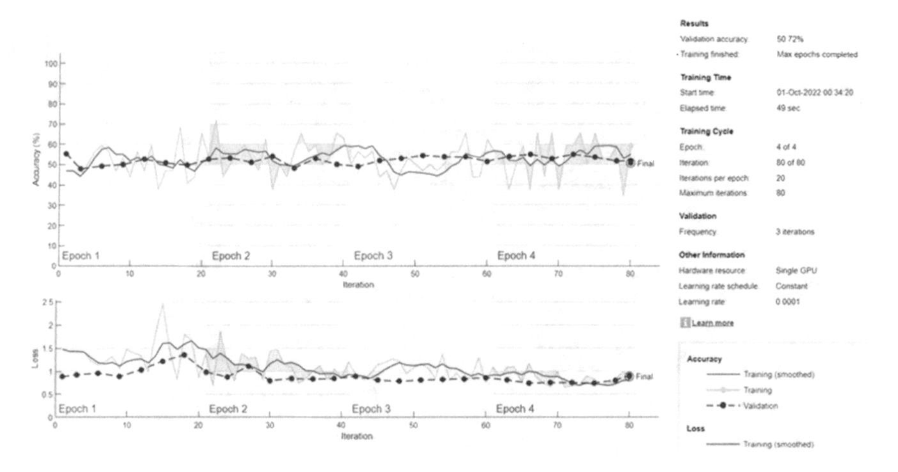

Fig. 19 Classification accuracy of single modalities for func Neuro imaging (50.72%)

after expanding the number of highlights, comes to its most extreme classification precision at a certain point, and at long last slips.

(d-1) **Classification analysis**:

The classification applies on classification models and multi-modal data fusion technique as the following as shown AlexNet Construction in Fig. 17. Figures 18 and 19 show the classification training using pre-trained AlexNet to create a classification neural network for Neuro imagining into two types. Figures 20 and 21 presents the training classification for two derivation descriptive data about MRI of Neuro imagining to improve the classification results.

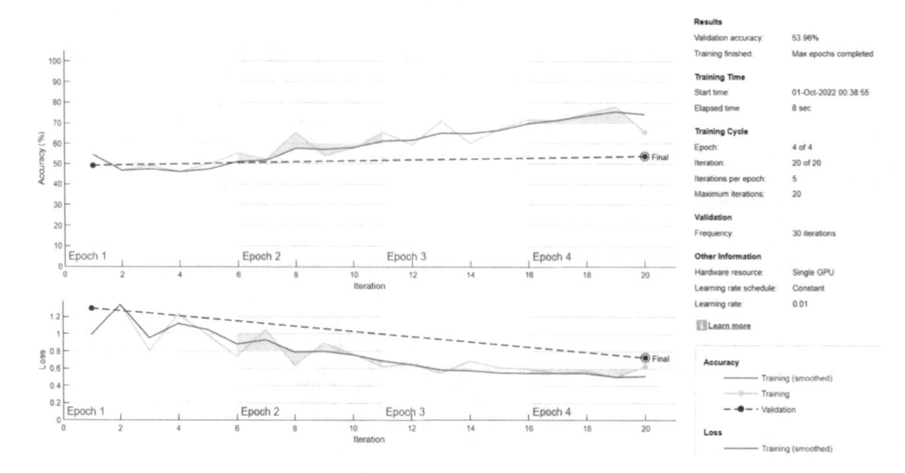

Fig. 20 Fusion classification accuracy of hybrid modalities using concatenation fusion for multi-modal of Neuro imaging (53.90%)

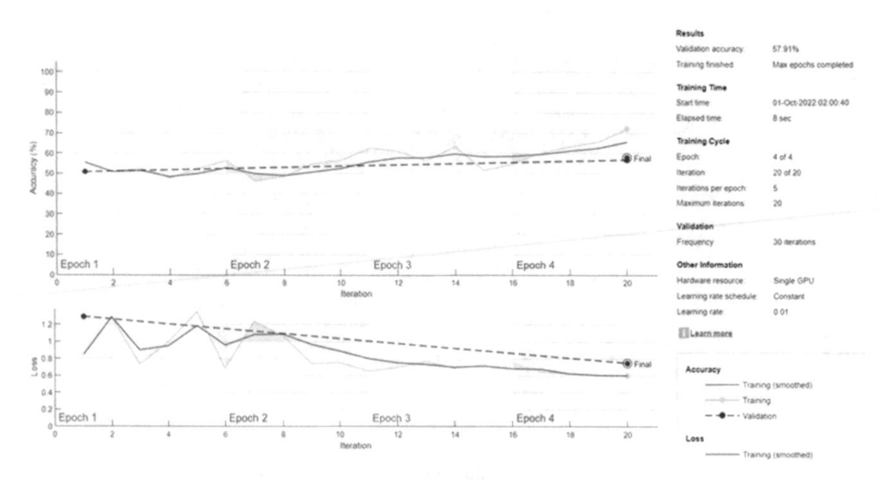

Fig. 21 Fusion classification accuracy of hybrid modalities using Dempster-Shafer fusion for multi-modal of Neuro imaging (57.91%)

(d-2) **Fusion analysis**:

Figure 20 shows the Fusion Classification accuracy results of hybrid modalities using concatenation fusion for multi-modal of Neuro imaging achieves to 53.98%. Concatenation fusion applies 4 numbers of epochs, 5 iterations per epoch, and the frequency iterations are 30 iterations.

Figure 21 shows the Fusion Classification accuracy results of Hybrid modalities using Dempster-Shafer fusion for Multi-Modal of Neuro Imaging achieves to 57.91%. Figure 22 presents the sample data of fused multi-modal Neuro imaging

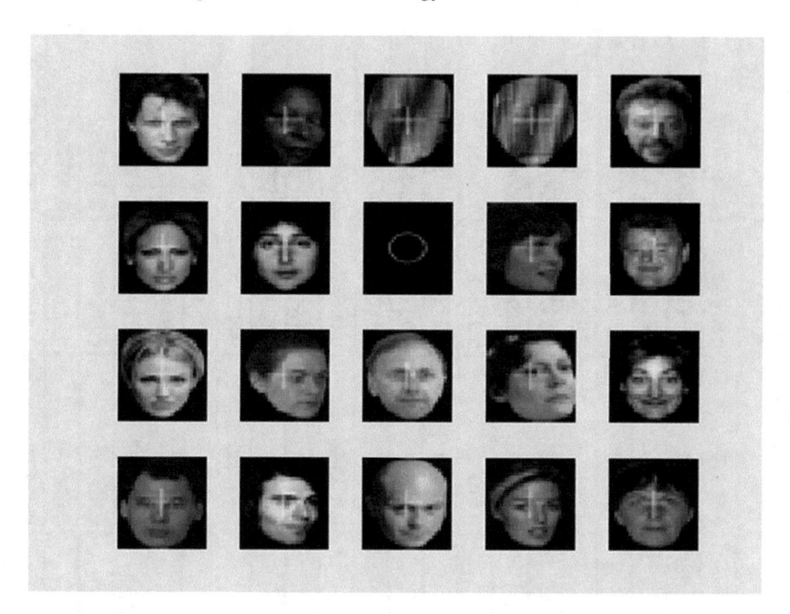

Fig. 22 A sample data of fused multi-modal Neuro imaging data based on three datasets based on 4 inputs

data based on three datasets based on 4 inputs. Dempster-Shafer late fusion applies 4 numbers of epochs, 5 iterations per epoch, and the frequency iterations are 30 iterations.

Bayesian optimizer refers to a sequential design strategy for global optimization of black-box functions that does not assume any functional forms. It is usually employed to optimize expensive-to-evaluate functions. Bayesian optimization methods are efficient because they select hyperparameters in an informed manner (Table 1). Figure 23: The performance of Bayesian optimizer with tracing to achieve the best fit accuracy.

MaxObjectiveEvaluations of 30 reached. Total function evaluations: 30. Total elapsed time: 2962.9786 s. And Total objective function evaluation time: 2954.4835. Best observed feasible point as shown in Table 2, when section depth is 3, initial learn rate is 0.024679, the momentum is 0.8148, and L2Reulaization achieves to 0.0073691.

Observed objective function value = 0.42086. Estimated objective function value = 0.45761. And Function evaluation time = **93.4731**.

Best estimated feasible point (according to models) as shown in Table 3.

Estimated objective function value = 0.45726 and the estimated function evaluation time = **98.4237**. The valError = 0.4209.

The multimodal comes about uncovered a considerable improvement of classification execution characteristics.

Table 1 Bayesian optimizer for searching the best fit point in multi-modal fusion

Iter	Eval result	Objective runtime	Objective	Best So Far (observed)	Best So Far (estim.)	SectionDepth	InitialLearnRate	Momentum	L2Regularization
1	**Best**	**0.44964**	**96.86**	**0.44964**	**0.44964**	**2**	**0.06337**	**0.97664**	**1.9357e-10**
2	Accept	0.48201	94.905	0.44964	0.45137	3	0.07151	0.94721	0.00033893
3	Accept	0.48201	99.996	0.44964	0.44964	1	0.085746	0.81472	0.00065955
4	Accept	0.48561	93.821	0.44964	0.44964	3	0.068404	0.8166	1.8847e-08
5	Accept	0.5	97.926	0.44964	0.44999	1	0.7343	0.97153	6.8129e-07
6	Accept	0.47482	96.436	0.44964	0.45192	2	0.19739	0.82909	0.0014356
7	Accept	0.4964	103.01	0.44964	0.4815	2	0.06092	0.96774	2.0075e-10
8	Accept	0.4964	98.22	0.44964	0.48336	2	0.010721	0.97911	1.3775e-10
9	Accept	0.46403	100.07	0.44964	0.48121	2	0.12079	0.97869	1.4198e-10
10	Accept	0.51439	103.02	0.44964	0.48453	2	0.11849	0.97785	2.6118e-10
11	Accept	0.48201	100.38	0.44964	0.4843	3	0.71926	0.81241	0.0059288
12	**Best**	**0.44604**	**100.19**	**0.44604**	**0.48111**	**3**	**0.031888**	**0.80647**	**0.0098563**
13	**Best**	**0.42446**	**98.834**	**0.42446**	**0.46982**	**3**	**0.02275**	**0.81614**	**0.0059173**
14	Accept	0.5	97.223	0.42446	0.47842	3	0.012027	0.83844	0.0037779
15	Accept	0.43165	94.244	0.42446	0.42447	3	0.031285	0.80303	0.0091228
16	**Best**	**0.42086**	**93.473**	**0.42086**	**0.43451**	**3**	**0.024679**	**0.81848**	**0.0073691**
17	Accept	0.48921	94.591	0.42086	0.42087	2	0.024267	0.94823	0.0071497
18	Accept	0.48561	102.14	0.42086	0.42087	3	0.034057	0.81459	0.0060507
19	Accept	0.44604	102.18	0.42086	0.42087	3	0.024336	0.82268	0.0072797
20	Accept	0.43885	102.19	0.42086	0.42087	3	0.024409	0.8133	0.0051977
21	Accept	0.47842	103.31	0.42086	0.42087	2	0.020507	0.81969	0.0057434

(continued)

Table 1 (continued)

Iter	Eval result	Objective runtime	Objective	Best So Far (observed)	Best So Far (estim.)	SectionDepth	InitialLearnRate	Momentum	L2Regularization
22	Accept	0.46763	101.4	0.42086	0.42087	3	0.018213	0.82203	0.0078554
23	Accept	0.47842	99.812	0.42086	0.42087	3	0.019766	0.81929	0.0043074
24	Accept	0.48561	96.195	0.42086	0.42087	3	0.036795	0.80296	0.0062318
25	Accept	0.4964	101.44	0.42086	0.4588	3	0.025196	0.8163	0.0078974
26	Accept	0.46763	103.64	0.42086	0.45706	3	0.17638	0.92481	1.6287e-05
27	Accept	0.46043	98.181	0.42086	0.42087	2	0.082543	0.91789	1.4675e-10
28	Accept	0.48561	95.359	0.42086	0.45699	3	0.024601	0.97721	6.606e-08
29	Accept	0.44964	93.808	0.42086	0.45715	2	0.20293	0.82469	1.0247e-10
30	Accept	0.47482	91.637	0.42086	0.45726	2	0.32953	0.80546	1.104e-10

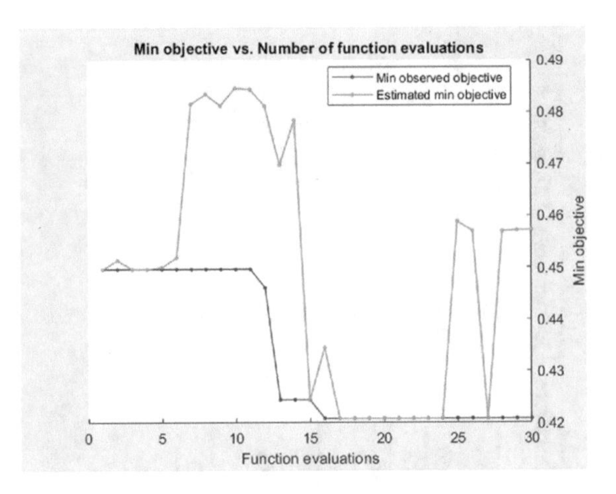

Fig. 23 The performance of Bayesian optimizer with tracing to achieve the best fit accuracy

Table 2 Bayesian optimizer shows the best fit point in multi-modal fusion

SectionDepth	InitialLearnRate	Momentum	L2Regularization
3	0.024679	0.81848	0.0073691

Table 3 Bayesian optimizer shows the best fit point in multi-modal fusion

SectionDepth	InitialLearnRate	Momentum	L2Regularization
3	0.024336	0.82268	0.0072797

8 Conclusion and Future Work

Metaverse technology has gotten to be the progressively prevalent shape of collaboration inside virtual universes. Such stages give clients with the capacity to construct virtual universes that can reenact real-life encounters through diverse social exercises. The concatenation early fusion experiment for hybrid modalities improves the accuracy results with 6.5–3.5%. The Dempster-shafer technique late fusion experiment for hybrid modalities improves the accuracy results with 9.5 and 7.5%. The Bayesian optimization improves the fusion of classification result for brain analysis with best fit point to 80 and 81%. The optimized prepare of selecting highlights to extend classification execution was based on investigating three properties of the combined highlights, counting diminishing repetition, increasing relevance and expanding complementarity. The future work is applied on larger dataset on multi modal brain signals.

References

1. Sang-Min, P., Young-Gab, K.: A metaverse: taxonomy, components, applications, and open challenge. IEEE Access **10** (2022)
2. Ashraf, D., Aboul Ella, H.: Fantasy magical life: opportunities, applications, and challenges in metaverses. J. Syst. Manag. Sci. **12**(2), 405–430 (2022)
3. Top Metaverse Statistics in 2022 available online, https://www.amraandelma.com/metaverse-statistics/
4. https://www.precedenceresearch.com/metaverse-in-healthcare-market
5. Alsharif, H., Mohamad, A., Amani, Y., Majdi, O.: A technology acceptance model survey of the metaverse prospects. AI J. **3**, 185–302 (2022)
6. Yuntao, W., Zhou, S., Ning, Z., Rui, X., Dongxiao, L., Tom, H.L., Xuemin, S.: A survey on metaverse: fundamentals, security, and privacy (2022). arXiv:2203.02662[cs.CR]
7. Pew Research Center (2022). https://www.pewresearch.org/internet/2022/06/30/the-Metaverse-in-2040/
8. Doaa, M.E., Aboul Ella, H., Ehab, E.H., Walaa, M.E.H.: E-quarantine: a smart health system for monitoring coronavirus patients for remotely quarantine (2020). arXiv preprint arXiv:2005.04187
9. The metaverse: a new challenge for the healthcare system: a scoping review. J. Funct. Morphol. Kinesiol. **7**(63) (2022)
10. Judith, N.N., Cosmas, I.N., Gaabriel, C., Dong-Seong, K.: The metaverse: a new challenge for the healthcare system: a scoping review. J. Funct. Morphol. Kinesiol. Wiley, IET Intelligent Transport Systems **7**(63) (2022)
11. Yuntao, W. et al.: A Survey on Metaverse: Fundamentals, Security, and Privacy, DeepAI. (2022)
12. Ling, Z.: The Metaverse: Concepts and Issues for Congress, Congressional Research Service. (2022)
13. Panagiotis, T., et al.: An artificial intelligence-based collaboration approach in industrial IoT manufacturing: key concepts, architectural extensions and potential applications. Sensors **20**(19) (2020)
14. Rabi, N.B., Kajaree, D.: A survey on machine learning: concept, algorithms and applications. Int. J. Innov. Res. Comput. Commun. Eng. **2**(2) (2017)
15. Dhanesh, R., Graham, W.T.: Deep multimodal learning: a survey on recent advances and trends. IEEE Signal Process. Mag. **34**(6), 96–108 (2017)
16. Minrui, X., Wei, C.N., Yang, B., et al.: A full dive into realizing the edge-enabled metaverse: visions, enabling technologies, and challenges (2022). arXiv:2203.05471v2[cs.NI]
17. Clement, J.: The statistics of investment industry fields of metaverse. https://www.statista.com/statistics/1302091/global-business-sectors-investing-in-the-metaverse/ (2022)
18. Doaa, M.E., Aboul, E.H., Ehab, E.H.: The adaptive smart environment multi-modal system. J. Syst. Manag. Sci. **12**(3), 1–45 (2022)
19. Lik-Hang, L., Tristan, B., Pengyan, Z., et al.: All one needs to know about metaverse: a complete survey on technological singularity, virtual ecosystem, and research agenda. J. Latex Class Files **14** (2021)
20. Ma, Y., et al.: Multimodality in meta-learning: a comprehensive survey (2022). arXiv:2109.13576v2[cs.LG]
21. Aleksadar, J., Aleksandar, M.: VoRtex Metaverse Platform for Gamified Collaborative Learning, Electronics, vol.11(3) (2022)
22. Alexei B., et al.: data2vec: a general framework for self-supervised learning in speech, vision and language. In: Proceedings of the 39 the International Conference on Machine Learning, Baltimore, Maryland, USA, PMLR (2022)
23. Rohit G., et al., Omnivore: a single model for many visual modalities (2022). arXiv: 2201.08377[cs.CV]
24. Amanpreet, S., et al.: FLAVA: a foundational model spanning dozens of multimodal tasks (2022). arXiv: 2112.04482[cs.CV]

25. Thien, H., Quoc-Viet, Xuan-Qui, P., Artificial intelligence for the metaverse: a survey (2022). arXiv: [2202.10336v1][cs.CY]
26. Burhanettin, U., Tarık, S.: A new age in health: metaverse Gevher Nesibe. J. Med. Health Sci. **7**(18), 93–102 (2022)
27. Zaheer, A., Ayyoob, S., Simon, E.B., David, S.J., John, K.: The metaverse as a virtual form of smart cities: opportunities and challenges for environmental. Econ. Soc. Sustain. Urban Futures Smart Cities **5**, 771–801 (2022)
28. Nannan, X., Juan, C., Filipe, G., Marc, R., Juho: The challenges of entering the metaverse: an experiment on the effect of extended reality on workload. Inf. Syst. Front. (2022)
29. Huansheng, N., Hang, W., Yujia, L., Wenxi, W.: A survey on metaverse: the state-of-the-art, technologies, applications, and challenges (2021). arXiv:2111.09673[cs.CY]
30. Dataset available online, https://openfmri.org/dataset/ds000117/
31. Daniel, G.W., Richard, N.H.: A multi-subject, multi-modal human neuroimaging dataset. Sci. Data (2015)
32. Anter, A.M., Moemen, Y.S., Darwish, A., Hassanien, A.E.: Multi-target QSAR modelling of chemo-genomic data analysis based on extreme learning machine. J. Knowl. Based Syst., Elsevier **188**, 104977 (2020). https://doi.org/10.1016/j.knosys.2019.104977

Metaverse and Its Impact on Climate Change

Palak, Sangeeta, Preeti Gulia, Nasib Singh Gill, and Jyotir Moy Chatterjee

Abstract A novel idea called the metaverse combines various technologies to offer a lifelike experience in a virtual setting. To realize its full potential, the idea must still be refined. People's opinions of the phrase are conflicted. Whether the metaverse is a beneficial thing or whether it will harm our future has been heavily disputed. Many critics think it might cause more harm than good, while many regard it as the next logical step for humanity after the internet. The impact of the metaverse on the climate is just one of many worries. The enormous computer power needed to implement artificial intelligence (AI) techniques could increase carbon emissions. A metaverse may also decrease the need for physical travel, saving a significant amount of energy and resources and lowering fuel use and the emission of toxic gases. This article is a study of various technologies used to provide an immersive experience in the virtual world and their overall contribution to environmental climate change.

Keywords Metaverse · Multi-technologies · Virtual reality · Climate change · Carbon footprint

Palak
Department of Computer Science, Government College, Narnaul, India
e-mail: palak.aug6@gmail.com

Sangeeta · P. Gulia · N. S. Gill
Department of Computer Science and Applications, Maharshi Dayanand University, Rohtak, India
e-mail: sangirao5228@gmail.com

P. Gulia
e-mail: preeti@mdurohtak.ac.in

N. S. Gill
e-mail: nasib.gill@mdurohtak.ac.in

J. M. Chatterjee (✉)
Department of IT, Lord Buddha Education Foundation, Kathmandu, Nepal
e-mail: jyotirchatterjee@gmail.com

A. E. Hassanien et al. (eds.), *The Future of Metaverse in the Virtual Era and Physical World*, Studies in Big Data 123, https://doi.org/10.1007/978-3-031-29132-6_12

1 Introduction

The metaverse is the next stage of interaction and is an ultramodern concept. Since Facebook changed its name to "Meta" in 2021, the term "metaverse" has gained some popularity, but in essence, it refers to the fusion of immersive digital technologies to provide a life-like experience in a digitally constructed world. Modern technologies like virtual reality (VR), augmented reality (AR), artificial intelligence (AI), blockchain, and cryptography will be used to simulate the physical world in the metaverse. Augmented reality (AR) is a more sustainable form of VR that adds digital elements to a real setting, rather than creating an entirely new environment from scratch. Although the first Metaverse has not yet been constructed, there are already several examples that may be used to illustrate the concept.

A brand-new class of social media platforms and Internet applications called Metaverse incorporates many cutting-edge technologies. It exhibits sociality, hyper spatiotemporally, and multi-technology traits [1]. The idea of the metaverse is still one that is still developing, and various individuals are adding to its significance in unique ways. Our climate emergency needs us to act like grownups; the metaverse encourages us to behave like children. Before the metaverse is fully optimized, there is still much work to be done. Companies are vying with one another to develop the operating system of the metaverse as well as metaverse platforms. There are numerous industries where the metaverse can show to be a benefit for giving long-lasting experiences with better resource management and service delivery. Whether the metaverse is a beneficial thing or whether it will have an adverse impact on our future has been heavily disputed. Many critics think it might cause more harm than good, while many regard it as the next logical step for humanity after the internet. The biggest and most dangerous global challenge currently facing humanity is climate change. Extreme weather events are increasing, global temperatures are rising, and many species are in danger of going extinct. The impact of the metaverse on the climate is just one of many worries. It is no secret that any virtual experience requires energy. Recently the demand for this resource has risen well beyond the amounts we've required prior. Data Quest reports that experts are concerned that the metaverse could cause a rise in greenhouse gas emissions because it uses a lot of energy to train AI modules [2].

The need for cloud gaming for VR may result in an increase in carbon emissions by 2030. Additionally, the need for high-resolution photographs will rise, necessitating yet more energy. We can only speculate as to how the Metaverse will attain net-zero emissions as long as it is still in the early stages of development. We currently have no way of knowing if the Metaverse will be a success and overcome its uncertain environmental prospects.

2 Related Work

The concept of the metaverse is quite new and very less explored. Although the term "Metaverse" was coined long back in 1992 in **Neal Stephenson's science-fiction novel "Snow Crash"**[3, 4], but this road is still very less travelled. It is quite fascinating for AI and VR researchers. Even though we are still only on the doorstep of a virtual universe, setting up the Metaverse could be both challenging and full of opportunity. H. Duan et. Al consider metaverse opportunistic for students and implemented a blockchain-driven university campus prototype. As per their considerations, the three-layer architecture of metaverse is proposed which is a combination of the physical world, virtual world, and their interaction [5]. S. Mystakidis also described various prospects of a metaverse in the field of education. The author identifies various challenges related to metaverse implementation including health issues, ethics, data privacy, etc., and highlighted various application domains [6]. A similar threat is the concept of surrogate nature in which the human being will be far away from real nature in the metaverse. M. C. Rillig et. Al recognized such risks including energy consumption and biased nature representation [7]. A. Plechatá et. Al experimented to influence the dietary habits and food choices of 122 volunteers and observed a moderate-large decline in dietary carbon footprint one week after the intervention. The authors concluded that a complex VR intervention focused on visualizing the future consequences of individual food choices can promote positive change in sustainable eating habits [8]. The study of the literature reveals that we can only speculate as to how the Metaverse will attain net-zero emissions as long as it is still in the early stages of development.

3 Metaverse Key Technologies

The Metaverse is built on the convergence of augmented reality (AR), virtual reality (VR), and Artificial intelligence (AI) technologies, which enable multimodal interactions with digital items, virtual environments, and people. As a result, the Metaverse is a web of networked immersive experiences and social in multiuser persistent platforms. It requires a collaboration of many technologies working together in a seamless manner that the user is unaware of due to the high level of abstraction. Depending upon the services provided in the metaverse, various other technologies including blockchain, cryptocurrency, 5G, IoT, etc. as shown in Fig. 1 are involved in making the project successful. A small-scale version of the Metaverse might already exist with blockchain technology and NFTs, or non-fungible tokens. Blockchain is a highly secure digital transaction system that makes it difficult for users to access or change information in the system. NFTs are tokens that represent anything with a unique identity, including avatars and gaming items. The Metaverse could use these technologies to create a more streamlined and secure augmented reality.

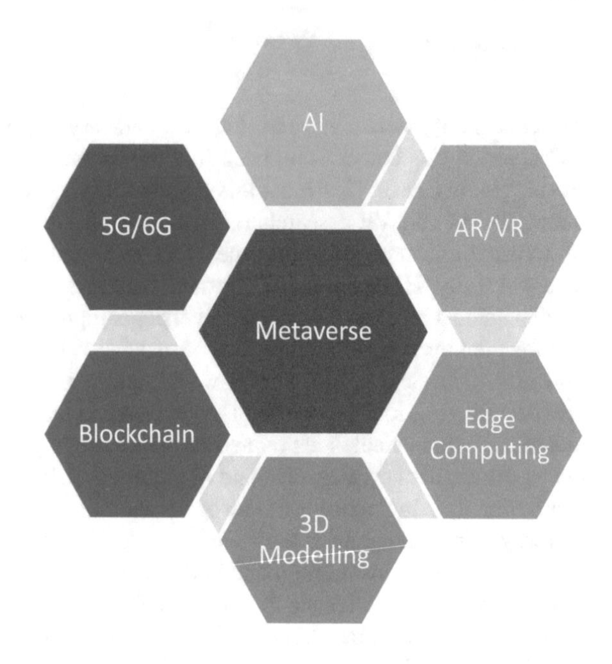

Fig. 1 Technologies Powering the Metaverse

4 Metaverse Challenges

Simultaneously, the dangers related to these ventures are challenging to survey as the desire to encounter reality may be more grounded than being taken part in virtuality [9]. These challenges offer tremendous opportunities ahead for consumers and companies alike [10].

a. *Identity*

One serious problem lies in proving your identity as bots can easily mimic your style, data, personality, and whole identity. You will need different verification methods like facial scans, retina scans, and voice recognition for authentication [11].

b. *Health-related issues*

Many research studies have explored the connections between children's use of electronic devices and mental health, and the results are clear: As use increases, so does the risk of mental health problems including depression, anxiety, ADHD, mood disorders, and suicidality. Teens who use electronic devices for more than two hours per day report significantly more mental health symptoms, increased psychological distress, and more suicidal ideation. Children using devices for more than 2 hours per day have an increased risk of depression, and that risk rises as screen time increases.

c. *Privacy and security*

A wide range of security breaches and privacy invasions may arise in the metaverse from the management of massive data streams, pervasive user profiling activities, and unfair outcomes of AI algorithms, to the safety of physical infrastructures and human bodies. There have been incidents of emerging technologies, such as the hijacking of wearable devices or cloud storage, theft of virtual currencies, and the misconduct of AI to produce fake news, hackers can exploit system vulnerabilities and compromise devices as entry points to invade real-world equipment such as household appliances to threaten personal safety, and even threaten critical infrastructures [12].

d. *Legal aspects*

As with any groundbreaking technological development, the metaverse will raise novel and complex legal issues [13]. Possible legal implications of the metaverse are around data protection, marketplace transactions, virtual assault, etc. Questions about intellectual property are also highly relevant.

e. *Network and Infrastructure*

Major infrastructure investment by tech companies is needed to reach this market potential. Issues like low latency, network bandwidth, image resolution, cloud storage, consumer devices, etc. need to be dealt with for a seamless experience.

5 Impact on Climate Change

The Metaverse can revolutionize and provide users with amazing experiences because of its vast application domain. Tech professionals from all around the world have been working diligently and coming up with innovative solutions to rework the original Metaverse concept. The world can be altered by the metaverse. However, it is debatable whether or not these modifications will be sustainable. Numerous debates have been made on this topic. So, we have categorized the impacts of the metaverse into two classes: positive impacts and negative impacts. The detailed description is given below.

a. ***Positive Impacts***

This section describes the positive aspects of the metaverse. The metaverse has the potential to become a continuous extension of people's lives in a parallel virtual environment where they can choose to live, work, and play. The most straightforward environmental benefit of the Metaverse is that it allows physical events, constructions, activities, and products to take virtual forms.

- *Reduced physical travel*

It would be simpler for political and corporate giants to hold conferences online, saving them both time and astronomical costs for their security and travel. The main advantage for the environment is that it will require a lot less human travel, which will lessen congestion, mishaps, pollution, and ultimately global warming. The Metaverse has the potential to significantly reduce local and worldwide transport emissions if it is successful.

- *Equality*

In the metaverse, social boundaries are not entirely significant. The metaverse can also be referred to as the "World with no discrimination". The anticipated metaverse should be a more physical and direct society with diminished notions of race, gender, and even physical impairment, which would be extremely advantageous for civilization [5]. Equality enhances opportunities and makes the world a better place to live. The basis for everyone's ability to enjoy long, healthy, and fulfilled lives is equality. Promoting gender equality improves everyone's quality of life by eradicating prejudice, damaging stereotypes, and violence based on gender.

- *Improved health sector facilities*

Metaverse can take the health sector and surgical procedures to a whole new level. The metaverse will give new, cutting-edge approaches to maintaining our health. There is already virtual reality counseling accessible, and therapists are employing VR headsets to administer exposure treatment to patients so they can confront their fears in a secure, controlled setting. Additionally, some surgical procedures are guided by augmented reality technology where surgeons and digital twins can be used to practice surgeries.

- *Improved weather forecasting*

AI and Machine learning-based technology provide better weather forecasting tools to help scientists analyze and plan accordingly [14, 15]. Weather forecasts are a major benefit for society and sustainable development. To reduce weather-related losses and increase societal advantages, such as the protection of life and property, public health and safety, and support for economic development and quality of life, weather forecasting aims to give information that individuals and organizations may use. With the metaverse, weather prediction will become more immersive and help in government planning and policy-making.

b. ***Negative Impacts***

It's also not always obvious how and to what extent a particular growing technology may affect our environment, even though many people view expansion in the tech sector as a sign of societal advancement. However, despite the potential for good, the Metaverse may also have unfavorable effects on society and the climate.

- ***Increased Energy Requirement and Carbon Emission***

Because of the recent explosion in AI and the increasing availability of data and processing power, the demand for energy consumption is rising exponentially. Metaverse combines many technologies and for each service layer, the need for energy should be optimized. "According to a recent study, just one AI model's training may produce 626,000 pounds of carbon dioxide, more than five times the amount of greenhouse gases generated by a car throughout its lifespan [16]". Moreover, energy consumption by VR/AR headsets, networking devices, and other gadgets involved also contribute to greenhouse gas emissions.

Data centers and Clouds

A "superabundance of cloud-streamed data" is required by metaverse to present persistent better real-world like experiences in the virtual environment. Very high computing power is required for the generation of these data. This will lead to the rise in compute-intensive, fast and efficient blockchain transactions to govern metaverse commerce. As the metaverse is an amalgamation of many technologies like AI, Virtual Reality, and blockchain, all these technologies will demand more energy consumption and hence eventually affect carbon emissions. The amount of carbon emission will be largely determined by the efficiency of the data centers and power sources. According to a 2020 Greening The Beast study, "high-end gamers, which have the hardware required for state-of-the-art VR, will spend as much as $2,200 over five years on electricity and pump as much as 2,000 pounds of carbon emissions into the atmosphere each year." (Table 1).

The various elements of the metaverse will be heavily dependent on costlier servers. It is very difficult to determine the actual environmental impact levied by individual data centers. One study in this context claims that in 2015, data centers contributed to about 2% of total global greenhouse gas emissions which is approximately equal to the same amount as produced by the whole aviation industry. The researchers at the Lawrence Berkeley National Laboratory claimed that "the additional energy used in cloud gaming can cause annual electricity use to rise 40 to 60% for desktops, 120 to 300% for laptops, 30 to 200% for consoles, and 130 to 260% for streaming devices in 2016." In the current scenario of the increased domain of metaverse, we can expect an abrupt increase in energy consumption (Table 2).

Moreover, a study by researchers at the U.K.'s University of Bristol also claims that "30% of gamers using 720p or 1080p devices were to transition to cloud gaming by 2030, it'd cause a 29.9% increase in carbon emissions. If 90% of gamers moved to the cloud, it'd increase gaming's overall carbon emissions by 112%." (Fig. 2).

Table 1 Expected power consumption and carbon emission in high-end gamers in 2020

User	Estimated power consumption of 5 years	Estimated carbon emission per year
High-end gamers (having hardware for state-of-the-art VR)	$2,200	2,000 pounds

Table 2 Rise in electricity consumption by cloud gaming in 2016

Some components for cloud gaming	A rise in additional annual electricity use
Desktops	40 to 60%
Laptops	120 to 300%
Consoles	30 to 200%
Streaming devices	130 to 260%

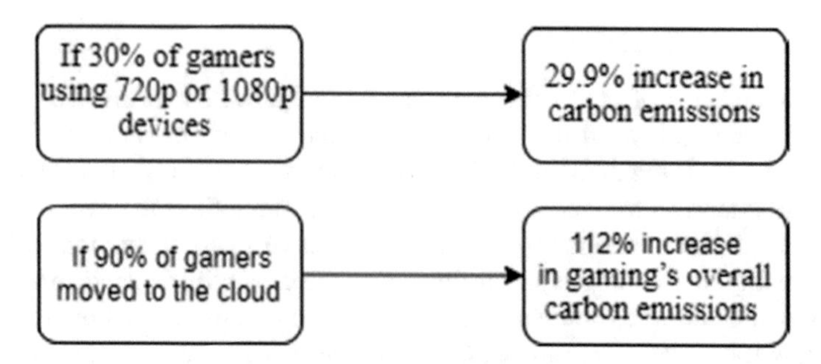

Fig. 2 Expected increase in carbon emission by cloud gaming in 2030

As metaverse engrosses the mass generation of decentralized data and this data needs to be stored in the cloud. Hence, more data centers are **required** to meet these growing needs, resulting in increased consumption of electricity. It also drives the shift of local data centers to the cloud. This shift may positively affect the environment.

Though the shift to cloud services emerged as a beneficial move but the energy requirements and consumption by data centers are also increasing. These days, IT cloud services account for about 2% of total global carbon emissions. The increased usage of digital services during the pandemic resulted in the release of 60 M tons emission of CO_2 in 2020 alone by Amazon. It could be foreseen that high-resolution imagery will require high data processing power in the metaverse. Moreover, as a massive number of users is brought into the universe from different platform providers, they will also significantly inflate these requirements.

According to Microsoft's lifecycle analysis, the Microsoft Cloud is about 22 to 93% more efficient than traditional local data centers in terms of energy consumption. To develop a sustainable and energy-efficient environment, cloud service providers are investing massive efforts in sustainable energy sources. Google claims that its data centers will be completely carbon-free by 2030 and Microsoft pledged to achieve this goal by 2025.

Though during the past decade, data center workloads from power consumption have been decoupled by their respective efficiency gains. From 2010 to 2020, data centers increased 9.4x and internet traffic 16.9x with only a 1.1x increase in power consumption. If cryptocurrency mining is not considered, only 1% of global electricity demand is contributed by data centers (Table 3).

Such efficiency gains are achieved with the use of hyper-scale data centers. These are specifically designed to restrain energy costs and invest in continual upgrades

Table 3 Affect of efficiency gains of data centers over power consumption

Year	Increase in internet traffic	The rise in data centers	Increase in power consumption
2010–2020	16.9x	9.4x	1.1x

and optimal sitting. The shift to Hyperscale represented a shift of 6% in 2015 to 45% of data center energy demand in 2021.

Some of the tech companies are meeting their electricity needs completely with renewable energy sources while others are also moving to "24x7 renewables", where the energy demand of each data center will be met by location-specific renewables at all times. These decarbonization achievements and impressive efficiency may be challenged by several other factors also. Data localization and privacy requirements may drive their companies to reinstate or keep their data centers. Moreover, for lifelike and compelling VR experiences low latency is demanded, which could shove data processing to the edge of networks, closer to users. Some data centers have to be in energy markets where acquiring renewables is not easy, or where operating conditions may affect the efficiencies or carbon-intensive back-up is required.

Training AI Models

As per Metaphysic's blog, "There will be markets for trading AI models to generate new content, much like users can purchase custom in-game items today. The combination of AI-generated content and virtual reality will allow for total immersion in alternative realities," But to train and run these types of generative AI systems, a lot of computation power is required. If we consider an example of OpenAI's DALL-E system, it comprises a text-writing AI system GPT-3, which is trained on pairs of text and images from the internet. 1,287 megawatts of energy is required for its training and consequently generates 552 metric tons of carbon dioxide emissions [17] (Table 4).

Some researchers at the University of Massachusetts said that 626,155 pounds of CO2 equivalent are emitted in training a single transformer AI model with 213M parameters [18]. As the platform providers try to provide enhanced user experience or provide relevant advertisements to each user, they continually invest in upgrading and developing the most accurate AI model. To build and test a final "paper-worthy model" approximately 4,789 models are trained over a six-month period, which releases more than 78,000 pounds of CO2 equivalent. This training process will be repeated continually for upgrading the commercial AI models in the metaverse and will significantly increase carbon emissions. It will be very dreadful for the environment.

Table 4 Power consumption and carbon emissions in training in different AI models

AI system	Carbon emission
Training GPT-3	552 metric tons of carbon dioxide emissions
Training a single transformer AI model with 213Â M parameters	626,155 pounds of CO2
In the training of 4,789 models over six months for building and testing a final â€œpaper-worthy model	78,000 pounds of CO2

Table 5 Carbon emission by single ethereum transaction

NFT	Carbon emission
One Ethereum Transaction	110 kg of CO2 equivalent to 42,734 VISA transactions or watching 18,253 h of Youtube

Environmental Impact of NFTs

In the metaverse, blockchain technologies will be of prime requirement. Blockchain offers non-fungible tokens (NFTs), which are unique for every audio, video, and photo of other media associated with them. NFTs come in form of digital creatures, music, artwork, avatars, and HTML code. It will also be in form of a plot of land in the virtual worlds. To enter and navigate among different worlds in the metaverse, the users can use authenticable and secure NFT avatars. As blockchain NFTs cannot be practically forged, they offer better security. Keeping track of ownership and trading of unique digital items, as offered by NFTs, will be an important component of the Metaverse.

According to a Digiconomist, "a single Ethereum transaction emits about 110 kg of CO2, equivalent to that of 42,734 VISA transactions or watching 18,253 hours of Youtube." When an NFT is transacted or minted, a new block is to be added to the blockchain, hence it would cost superfluously if creator studios are shifted to the Metaverse (Table 5).

Disposal of E-Waste

The risks posed by e-waste are real, and they are made worse by the unorganized sector, which frequently strips e-waste of its most beneficial components. Lethal substances like lead, cadmium, beryllium, mercury and brominated flame retardants are present in all electronic garbage. The likelihood of these hazardous compounds, contaminating the land, poisoning the air, and leaking into water bodies increases when gadgets and devices are disposed of illegally. The amount of worn and abandoned electronics is increasing along with the global demand for electronic devices. Every year, around 50 million tonnes of e-waste are produced, which more than the combined weight of all commercial aircraft is ever built. Instead with these things in mind, one can conclude that the Metaverse will be more detrimental to the environment than beneficial.

6 Conclusion

The metaverse, which will soon play a significant role in our lives, needs to be appropriate for users at all organizational and economic levels. This article describes various technologies involved in realizing the metaverse. The world can be altered by the metaverse. However, it is debatable whether or not these modifications will be sustainable. Numerous debates have been made on this topic. So, we have categorized the impacts of the metaverse into two classes: positive impacts and negative impacts. Various positive and negative impacts of metaverse for climate change are analyzed. Additionally, by 2030, all of the IT behemoths, including Meta, Google, Microsoft, Apple, and others, will transition to zero-net-emissions and sustainable computing, utilizing more renewable energy sources. We can only speculate as to how the Metaverse will attain net-zero emissions as long as it is still in the early stages of development. We currently have no way of knowing if the Metaverse will be a success and overcome its uncertain environmental prospects.

References

1. Ning, H., et al.: A survey on metaverse: the state-of-the-art, technologies, applications, and challenges (2021). http://arxiv.org/abs/2111.09673
2. Kwatra, M.: The Metaverse: what are the environmental impacts and future (2022). https://www.dqindia.com/the-metaverse-what-are-the-environmental-impacts-future/
3. Stephenson, N.: Snow Crash, 7th edn. Bantam Books, New York
4. Brown, D.: "What is the 'metaverse'? Facebook says it's the future of the Internet". Washington Post (2021). https://www.washingtonpost.com/technology/2021/08/30/what-is-the-metaverse/
5. Duan, H., Li, J., Fan, S., Lin, Z., Wu, X., Cai, W.: Metaverse for social good: a university campus prototype. MM 2021—Proc. 29th ACM International Conference Multimedia (2021), pp. 153–161. https://doi.org/10.1145/3474085.3479238
6. Mystakidis, S.: Metaverse. Encyclopedia **2**(1), 486–497 (2022). https://doi.org/10.3390/encyclopedia2010031
7. Rillig, M.C., et al.: Opportunities and risks of the 'metaverse' for biodiversity and the environment. Environ. Sci. Technol. **56**(8), 4721–4723 (2022). https://doi.org/10.1021/acs.est.2c01562
8. Plechatá, A., Morton, T., Perez-Cueto, F.J.A., Makransky, G.: Virtual reality intervention reduces dietary footprint: implications for environmental communication in the metaverse, no. 1, (2022). https://psyarxiv.com/3ta8d/%0Ahttps://psyarxiv.com/3ta8d/download?format=pdf
9. Hari Lal, S., M. K. K,: A study on virtual world of metaverse. Int. Res. J. Mod. Eng. Technol. Sci. **4**(05), 4535–4538 (2022)
10. Elnaj, S.: The challenges and opportunities with the metaverse (2022). https://www.forbes.com/sites/forbestechcouncil/2022/05/17/the-challenges-and-opportunities-with-the-metaverse/?sh=473f1445495f
11. Hazan, S.J.: Musing the metaverse. Herit. Digit. era, 95–104 (2010)
12. Wang, Y, et al.: A survey on metaverse: fundamentals, security, and privacy 1–31 (2022). http://arxiv.org/abs/2203.02662
13. Clifford Chance, "The metaverse: what are the legal implications?," no. February, pp. 1–7, 2022

14. Diaz, S.M.M.I., Combarro, E.F., et al.: Machine learning applied to weather forecasting. Springer **15**(2), 99999 (2017)
15. Haupt, S.E., Cowie, J., Linden, S., McCandless, T., Kosovic, B., Alessandrini, S.: Machine learning for applied weather prediction. In: Proceedings—IEEE 14th International Conference eScience, e-Science 2018, pp. 276–277 (2018). https://doi.org/10.1109/eScience.2018.00047
16. Strubell, E., Ganesh, A., McCallum, A.: Energy and policy considerations for deep learning in NLP. ACL 2019—Proceedings of the 57th Annual Meeting of the Association for Computational Linguistics, no. 1, pp. 3645–3650 (2020)
17. https://venturebeat.com/2022/01/26/the-environmental-impact-of-the-metaverse/
18. https://medium.com/geekculture/how-green-is-the-metaverse-the-two-sides-of-the-environmental-impact-of-the-metaverse-6a35913fd329

Cybersecurity in Metaverse

Blockchain Technology in Metaverse: Opportunities, Applications, and Open Problems

Mohamed Torky, Ashraf Darwish, and Aboul Ella Hassanien

Abstract The metaverse is the next technological revolution that will mirror our physical life into a virtual world. In this new world, various data patterns are being generated in large volumes, Avatars can perform various transactions Metaverse's major requirements are data reliability, integrity, security, transparency, and decentral integrity, security, transparency, and decentral are major requirements in Metaverse. Blockchain technology is one of the magic technologies able to optimally implement these requirements in the metaverse. This chapter discusses around why Blockchain is a key player in the metaverse. The chapter discusses the main characteristics of Blockchain and how they can be modeled in the metaverse. In addition, it discusses the differences between fungible and non-fungible tokens (NFTs), and how each token type can be processed using three types of Ethereum Request for Comments (ERC) protocols, ERC-20, ERC-721, and ERC-1155. After that, the chapter presents four real Blockchain applications in the metaverse. finally, the chapter is ended with a set of challenges and open problems around utilizing Blockchain technology in the metaverse.

Keywords Metaverse · Augmented reality · Virtual reality · Extended reality · Blockchain · Smart contract · Fungible tokens · Non-fungible tokens (NFTs) · Ethereum request for comments (ERC) · Self-sovereign identity (SSI)

M. Torky (✉)
Faculty of Artificial Intelligence, Egyptian Russian University (ERU), Scientific Research Group in Egypt (SRGE), Badr, Egypt
e-mail: mtorky86@gmail.com

A. Darwish
Faculty of Science, Helwan University, Scientific Research Group in Egypt (SRGE), Cairo, Egypt

A. E. Hassanien
Faculty of Computers and Artificial Intelligence, Cairo University, Scientific Research Group in Egypt (SRGE), Cairo, Egypt

A. E. Hassanien et al. (eds.), *The Future of Metaverse in the Virtual Era and Physical World*, Studies in Big Data 123, https://doi.org/10.1007/978-3-031-29132-6_13

1 Introduction

In the recent context, the concept of metaverse attracted the attention of the world and became a new era of technological progress by creating a virtual world that reflects all the details of real life. The end users can immerse in a virtual domain where they can do everything they do in their real life, such as meet friends, visit exciting places, buy objects and do the shopping and sell real estate. A set of promising 3D technologies such as Augmented Reality (AR), Virtual Reality (VR), Extended Reality (ER) [1] as well as Artificial Intelligence (AI) [2], and Blockchain are base technologies on which metaverse systems will work [3, 4]. Various leading companies started to invest in metaverse-based applications. For example, Meta, The tech giant formerly known as Facebook spent $10 billion on the metaverse in 2021, most of these investments have been paid on VR applications, headsets, and smart glasses. Those products are major to Mr. Zuckerberg's vision of the metaverse, where users would share virtual domains and experiences across various software and hardware platforms [5]. Microsoft, the giant of software already develop holograms and extended reality (XR) applications with its Microsoft Mesh platform, which integrate the physical world with virtual reality and augmented reality. Microsoft plans to use these holograms and virtual avatars in Microsoft Teams 2022. Moreover, The U.S. Army is currently cooperating with Microsoft on an augmented reality Hololens 2 headset for training soldiers to rehearse and fight in. Other companies invest in developing game applications in the metaverse with virtual capabilities such as building homes, working, and playing out scenarios, Roblox and Epic Games are common game platforms in the metaverse [5]. If VR, AR, and XR will design virtual objects, and avatars and shape the virtual space, Blockchain will process, manage, validate and secure all transactions between avatars and virtual objects and will do as a decentralized and transparent repository for a huge number of terabytes of virtual data patterns.

Experts agree and forecast that Blockchain, one of the promising technologies can open up an amazing virtual space, which will completely change the traditional ways of user communications and transactions. Without blockchain, the metaverse system is incomplete because of the various limitations of centralized data warehousing approaches. In addition, the open nature of the metaverse will result in a huge number of transactions and various patterns of communications, which require untraditional technology to manage, process, store, validate, and secure them. Moreover, the de-centrality feature of blockchain makes it an optimal technology to make the metaverse function globally. Hence, the blockchain-based metaverse provides access and performs various transactions to any digital space without the interference of centralized systems as in the recent internet applications, such as social networks and websites. In addition, blockchain will be the backbone of digital currency in the metaverse where a set of Blockchain protocols will manage fungible and non-fungible (NFTs) tokens, which are related to various and huge streams of metaverse transactions [6]. These advantages make Blockchain a base platform for building the financial system in the metaverse through a set of smart contracts that can work together as a middleware for processing fungible and NFTs tokens in the metaverse.

NFTs are created using a Blockchain protocol for each digital asset in the metaverse, these NFTs are unique and not replaceable, the opposite of fungible tokens like currencies (e.g. bitcoin, litcoin, ETH, etc.). A set of brands already started to use NFTs in various virtual applications, including Nike, Dolce & Gabbana, Adidas, and Coca-Cola [7]. In the future, when you buy a real-world object such as a car, home, or clothes, you might also gain ownership of a linked and corresponding NFT in the metaverse through a digital wallet (or crypto wallet), which will hold your digital assets and identity. Section 4 will discuss the NFTs concept in more detail.

The layout of this chapter can be organized as follows: Sect. 2 discusses the main features that make Blockchain a major technology and a key player in the metaverse. Section 3 discusses Blockchain protocols in Metaverse. Section 4 discusses fungible and non-fungible tokens in the metaverse. Section 5 provides some of the recent Blockchain applications in Metaverse. Section 6 discusses some of the challenges and open problems of the Blockchain-based metaverse. Section 7 summarizes the main concepts and approaches provided in this chapter.

2 The Importance of Blockchain Technology in Metaverse

Although there is still no standardized definition of the concept of the metaverse and the discipline itself is only partially developed in some projects such as the Google Blocks, metaverse Facebook Horizon, and some game applications, Metaverse can be imagined as any system consisting of hardware and software. The hardware devices include many controllers such as gloves, headsets, AR glasses, and bodysuits, which help the users to perform various tasks and do multiple transactions through the metaverse. in a metaverse environment, the more the user is equipped, the more the transactions and behaviors become better for the user. On the other hand, many experts in the industry have come to agree that metaverse software has to be built on blockchain technology [7]. Blockchain can work as a secure decentralized repository where metaverse-transactional data can be stored and shared between the users of the metaverse. It becomes clear that blockchain can meet the functional and non-functional requirements of the metaverse. Figure 1 summarizes the most important features and services of Blockchain technology, which make it a base software system in Metaverse. The characteristics of Blockchain in the metaverse can be categorized as follows:

2.1 Security

Metaverse will be able to store data measured in exabytes; this fact raises the question of secure storage, communications, and synchronization. although security techniques will most certainly be developed over time, for now, users and business

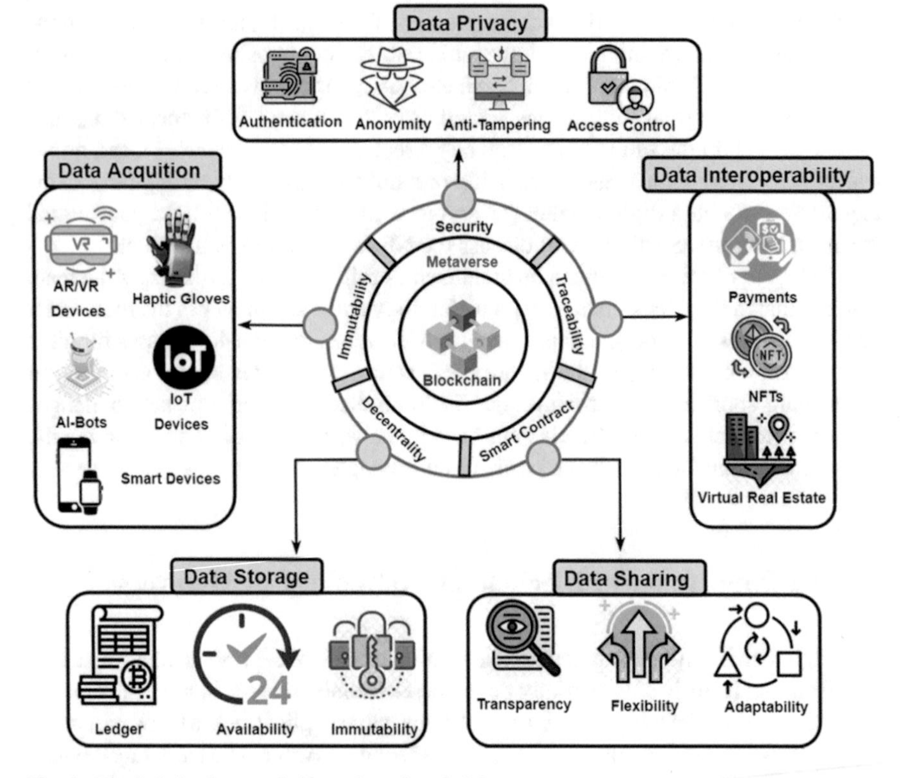

Fig. 1 Blockchain characteristics and services in Metaverse

administrators need to care about the following challenges when participating in the metaverse:

a. Preventing fraud and ensuring digital ownership of NFTs of virtual goods
b. Protection of avatars' identities and personal information
c. Protection of keys, seeds, and log-ins.
d. Money laundering risks with crypto assets in Metaverse.
e. Business cybersecurity in the metaverse.

For mitigating these challenges, blockchain technology can play a major role and provide untraditional solutions for securing metaverse data and transactions. For instance, blockchain uses symmetric keys, and Self-sovereign identity (SSI) [8] to implement digital identities that give metaverse-users control over the information they use to prove who they are in the metaverse platform. Figure 2 explains how SSIs are issued and verified in Metaverse. The most important difference between SSI and Federated Identity Model (FIM) is that SSI is not service provider-based. Instead, there is a direct secure peer-to-peer connection between the two parties. Neither peers nor any third parties can "control" the communication, this is true whether the parties are companies or people. Figure 3 summarizes the key differences between federated identity (FI) and SSI which will be applicable in the metaverse.

Blockchain can also enable metaverse users to use NFTs tokens to develop true and trusted ownership of in-game items such as avatars, weapons, lands, and buildings. This technology can prevent theft and fraud of metaverse transactions. Blockchain grants metaverse users control and real ownership of their metaverse assets. Using non-custodial-based wallets, the stockholders hold their cryptographic keys and seed phrases that allow them to access their crypto assets. One of the best blockchain-based services for file authenticity, e-signatures, and verification is *Acronis Cyber Notary Cloud* [9]. It enables customers and service providers to ensure the authenticity of

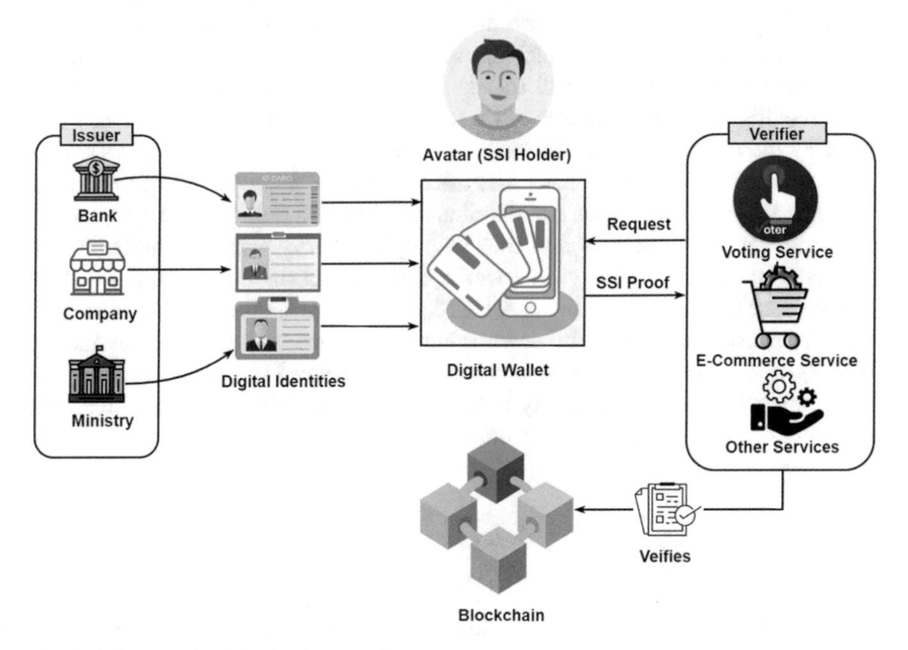

Fig. 2 Self-sovereign identity framework

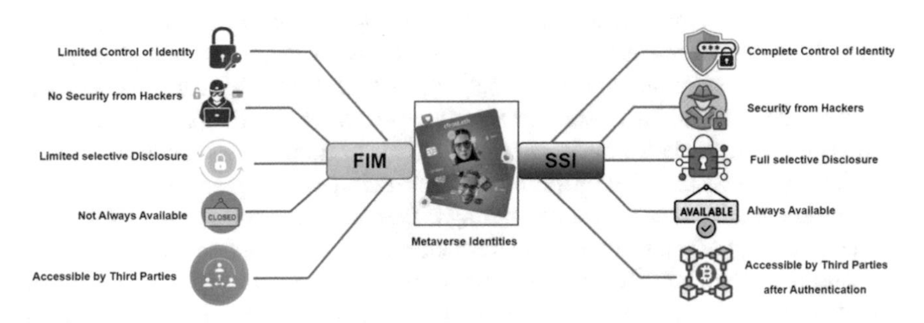

Fig. 3 Self-sovereign identity versus federated identity model

business-critical information, achieves regulatory transparency, and mitigates security risks in the metaverse. Hence, Blockchain can achieve all goals of multiverse data privacy such as authenticity, anonymity, anti-tampering, and access control.

2.2 Trust and Traceability

Trust in the metaverse is an important characteristic. It refers to the challenge of keeping the trust of both metaverse users and society in a virtual world where multiple technologies such as artificial intelligence, the Internet of Things (IoT), digital twins, and new interfaces can be integrated into a novel virtual world called "metaverse". Data penetrations, hacks, high jacking, data tampering, and avatar forgery, make the demand for transparency and traceability about the origin and type of data a major requirement in the metaverse. Trust and traceability are also important while verifying the metaverse's transactions within and across various virtual networks, interfaces, and applications. Trust and traceability can be achieved based on the zero-trust security approaches and distributed-ledger technologies (DLTs) to avoid metaverse breaches, ensure compliance with cybersecurity measures, and optimize transaction efficiency in this virtual world. Since, Blockchain technology work as a secure decentralized repository (i.e. DLTS) where independent nodes can interact without the need for a third party for verifying transactions and stored data, moreover, since it can store real-time data in a sequence of encrypted blocks, Blockchain can be considered the best technology able to achieve the requirements of trust and traceability as well as ensuring data quality in Metaverse.

2.3 Smart Contracts

Metaverse involves several economic domains, real estate, avatars, social networks, and many more entities. These make it mandatory to effectively regulate money transfer, economic communication, avatars' social connections, and other connections between participants and ecosystems within the metaverse. The known centralized systems are not able to effectively manage this huge amount of transactions due to security [10] and communication challenges. Therefore, the smart contract is a wonderful Blockchain-based solution that allows metaverse participants to develop, implement, ad regulate the basic rules of all kinds of transactions in the metaverse world. Hence, it ensures that operations such as trading, gaming, and avatar transactions are automatically carried out according to the prespecified rules. Figure 4 depicts the methodology of smart contracts in the metaverse. the transactions between metaverse users are being processed and verified by a smart contract through a built-in code that ensures applying a set of rules that validate a transaction in the metaverse. The verification process produces a novel block that store all details of these transactions and is securely added to the Blockchain.

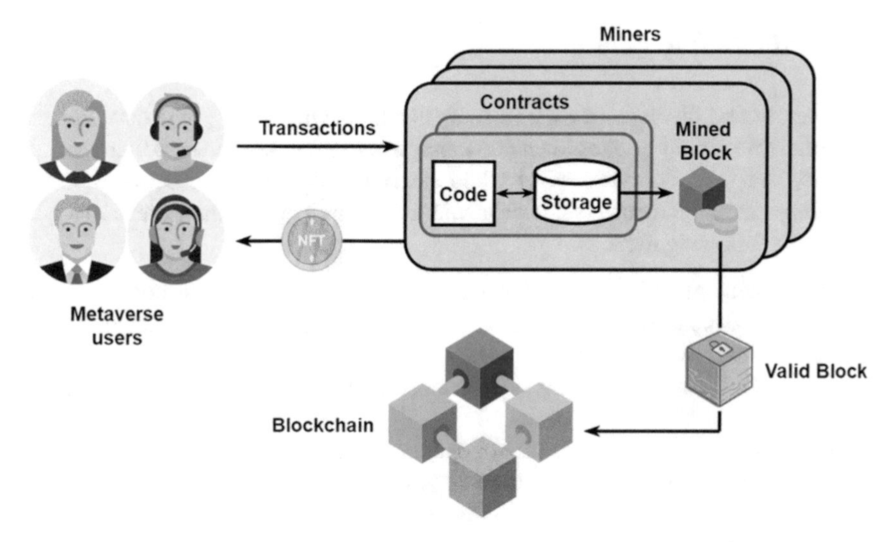

Fig. 4 Smart contract in Metaverse

2.4 Decentrality

In the metaverse, various systems and applications like traffic optimization, healthcare, gaming and entertainment, and E-commerce transactions will produce large volumes of real-time data. The way by which Data will be managed, verified, and shared becomes an important challenge. Moreover, the recent centralized techniques such as the client/server model will be ineffective and not suitable especially since the demand for verifying large volumes of real-time data increases in the metaverse. Therefore, a decentralized ecosystem based on blockchain technology allows thousands of avatars to synchronize their transactions in a decentralized fashion. Based on crypto tokens (e.g. NFTs), Blockchain can make the metaverse transactions more precise, and transparent while heterogeneous systems, applications, and avatars establish various transactions. In addition, the data owner will have full control over the shared information. The decentrality feature of Blockchain will decrease the time and money spent on verifying metaverse transactional and real-time data. However, as the number of metaverse users increases, the number of blocks must increase as well, this requires utilizing large amounts of computing resources. Therefore, the next generations of Blockchain have to consider this challenge to ensure effective data share and control in the metaverse.

2.5 Immutability

The variety and diversity of metaverse applications will result in the development of immutable virtual bridges to enable metaverse users to keep their avatars, virtual objects, and their 3D possessions while easily transferring them between virtual domains within the metaverse. Blockchain immutability [11] can achieve this goal due to the following features:

a. Blockchain provides efficient auditing services by creating a complete, and undoubted history of a transactional ledger. In addition, demonstrating that metaverse transactional information has not been altered and can mirror the real transactions in the real world is a significant strength for achieving the convergence between real and virtual worlds.
b. Blockchain ensures the integrity of data in the metaverse. The integrity of metaverse transactions can be approved by just inferring the hash code of a block in the blockchain. If there is any difference between block data and its related hash, it means that these transactions cannot be considered valid in the metaverse. This certainly, enables metaverse users and organizations to identify and verify data modification rapidly.

3 Fungible and Non-fungible Tokens in Metaverse

In the metaverse, there will be huge numbers of digital assets that can symbolize both internet collectibles and real-world objects. Thanks to blockchain, almost anything in pixels will be tokenized and turned into Non-Fungible Tokens (NFTs) [12]. NFTs is defined as a unique digital identifier that are impossible to be substituted, copied, or subdivided. NFTs are used to certify the ownership and authenticity of digital assets then, NFTs verifies are registered in a Blockchain. Therefore, NFTs will be a key player in the metaverse, which will be the base for enabling ownership and value of digital assets to be encoded and transferred across metaverse domains. Thanks to NFTs, The metaverse will reflect a hop from the internet of things (IoT) to the internet of values (IoV), the crucial enabler of the metaverse. Table 1 compares IoT and IoV as two technological enablers. On the other hand, fungible tokens are an economic concept that, represents objects that can be exchanged for other objects of the same type with no change in value. For example, one Bitcoin is equal to another Bitcoin, in brief words, fungible tokens refer to cryptocurrencies that will be exchanged in Metaverse. Figure 5 depicts fungible and non-fungible tokens (NFTs) and the processing of both of them using Blockchain.

Table 1 Internet of Things versus Internet of Values

Attribute	Internet of Things (IoT)	Internet of Values (IoV)
Connection	Connection is based on physical networks that connect various heterogeneous devices, organizations, businesses, governments, communities, etc.	Connection is based on virtual networks that connect avatars, virtual objects, NFTs-based systems, etc.
Data transferee	Data exchange and transactions are based on IoT protocols and the physical environments in which data is transferring	Data transfers in the form of asset values, such as cryptocurrencies, crypto assets based on blockchain protocols, and virtual domains in which data transfer
Unit of connection	Things: real-world devices, objects, animals, materials, etc.	Values: tokenized digital assets, and augmented sociality

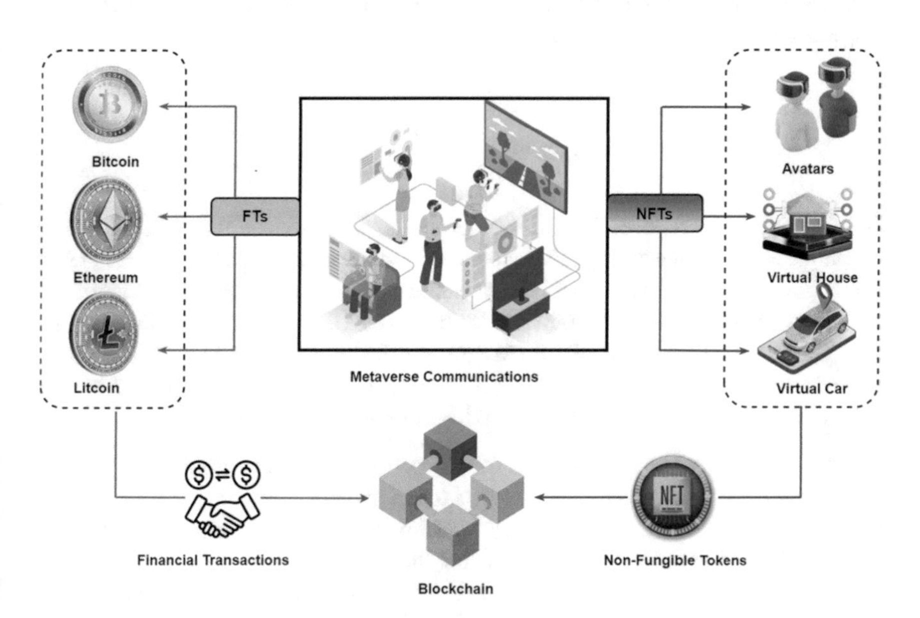

Fig. 5 Fungible and non-fungible tokens in metaverse

4 Blockchain Protocols in Metaverse

In the first generation of Blockchain technology, Bitcoin was working based on a set of complex protocols (e.g. Proof of Work (PoW) [13], Proof of Stake (PoS) [14], Delegated Proof of Stack (DPoS) [15]). These types of protocols were consuming large quantities of power, software, and hardware resources for solving a cryptographic puzzle or holding and staking tokens for mining Bitcoins [16, 17]. The open source Blockchain with programmable smart contracts is the next generation of Blockchain technology (e.g. Ethereum), which was developed by Vitalik Bueterin to address the limitations of the first generation of Blockchain. In the metaverse, Ethereum-based tokens can symbolize objects' values, and services, then avatars and virtual organizations can use these tokens as internal currencies to verify various types of a transaction within the metaverse. Recently, the Ethereum community develops some token standards called Ethereum Request for Comments (ERC) to create processes, and verify Ethereum tokens. In addition, it ensures the compatibility of the created tokens on Ethereum with various existing industrial ecosystem requirements. ERC standard is defined as a smart contract that specifies a set of rules and constraints that every Ethereum-based token must adhere to for enabling major operations such as transactions, token creation, transaction processing, token verification, spending, etc. ERC-20, ERC-721, and ERC-1155 appear as the three common ERC token standards that can be utilized for developing various decentralized applications on Ethereum across various domains in the metaverse [18]. The Ethereum community is uniquely responsible for approving and updating these token standards, which differ in specific functions and purposes.

4.1 ERC-20 Token Standards

ERC-20 was first proposed by Fabin Vogelstellar in 2015 as a way to standardize the fungible tokens (exchangeable tokens) within smart contracts on the Ethereum Blockchain. Fungible tokens can be represented virtually as digital currencies, such as Tether (USDT), USD Coin (USDC), Shiba Inu (SHIB), BNB (BNB), DAI Stablecoin (DAI), HEX (HEX), MAKER (MKR), and Bitfinex LEO (LEO) [19]. ERC-20 methodology is described through a set of functions, which govern how fungible tokens are created, processed, transferred, and verified as follows:

- *Total_Supply ():* The total number of tokens that will ever be issued
- *balance of ():* The account balance of a token owner's account
- *Transfer ():* Automatically executes transfers of a specified number of tokens to a specified address for transactions using the token
- *Transfer_From ():* Automatically executes transfers of a specified number of tokens from a specified address using the token
- *Approve ():* Allows a spender to withdraw a set number of tokens from a specified account, up to a specific amount
- *Allowance ():* Returns a set number of tokens from a spender to the owner
- *TransferEvent ():* An event triggered when a transfer is successful (an event)
- *Event Approval ():* A log of an approved event (an event).

Figure 6 depicts the methodology of ERC-20 for processing fungible tokens in Metaverse.

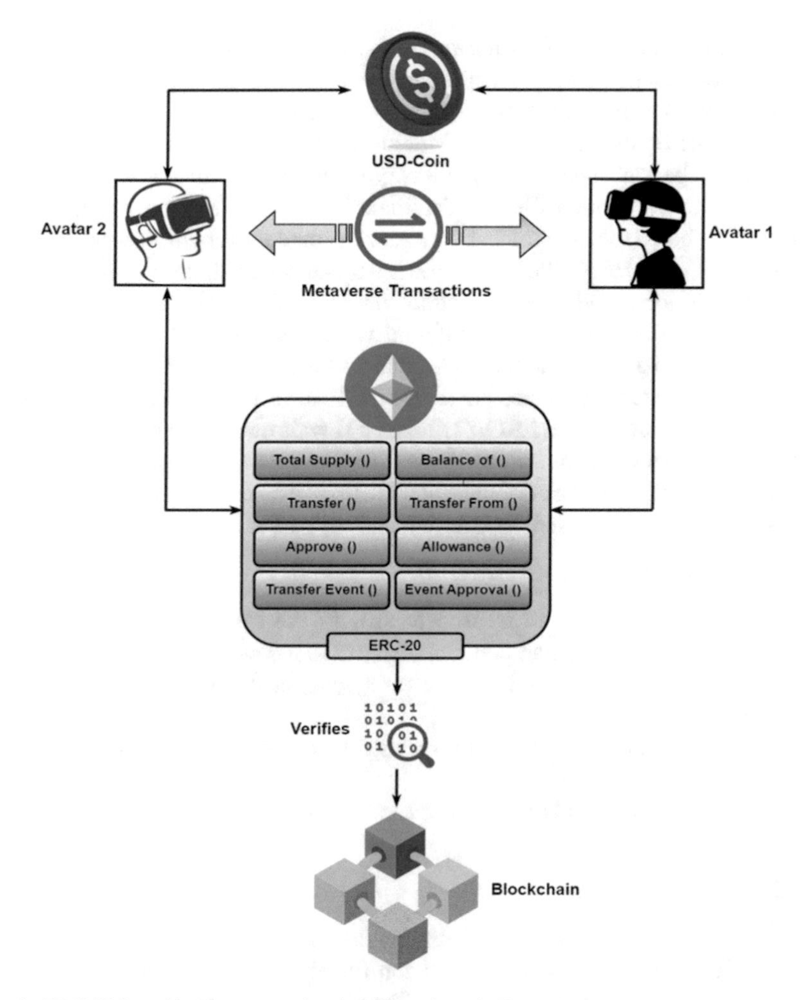

Fig. 6 ERC-20 functions for processing fungible tokens in the metaverse

4.2 ERC-721 Token Standards

The ERC-721 has been proposed by William Entriken, Dieter Shirley, Jacob Evans, and Nastassia Sachs in 2018 as a blockchain standard for creating, processing, and verifying non-fungible tokens (NFTs) [20]. In the metaverse, NFTs will be used to identify virtual objects (e.g. virtual cars, real estate, event tickets, Tweets and social media posts, In-game collectibles, etc.) or avatars uniquely. The digital representation of physical assets using NFTs in the metaverse will make a revolution and major change in crypto-currencies exchange. Recent finance systems consist of complex loan and trading systems for various asset kinds, ranging from winning objects in games to trading real estate. Using digital representations of physical assets, NFTs represent a big step forward in reforming the infrastructure of managing physical assets. NFTs also contain ownership data for easy recognition, specification, and transfer between metaverse users. NFTs holders can also add additional attributes or metadata related to the asset in NFTs. Moreover, since NFTs are one-of-a-kind, these tokens cannot be exchanged for any other cryptocurrency of equal value, where each NFT is associated with a different identifier. Therefore, NFTs are much different from ERC20-based tokens, which are a token standard for fungible tokens, meaning each token is interchangeable. In the ERC20 token standard, developers can create any number of tokens within one contract, but in the ERC721 token standard, each token within the contract holds a different value. The ERC721 protocol methodology is based on a set of common rules that all tokens can follow on the Ethereum network to generate the expected NFTs tokens. Token standards primarily undertake the following features about ERC-721 tokens: How is ownership decided? How are tokens created? How tokens are transferred? How tokens are burned? ERC-721 work is described through a set of attributes functions, and events, which govern how NFTs tokens are created, processed, transferred, and verified as follows [21]:

A. Functions similar to that of ERC-20

1. *Name:* This attribute is used to specify the token's name, which other applications and smart contracts can use to recognize it.
2. *Symbol:* This attribute is used as the symbol or shorthand name for the NFT token.
3. *Total_Supply ():* This function is used to specify the total supply of tokens on the Blockchain.
4. *Balance_of ():* This function Returns the number of NFTs that an address owns.

B. Ownership Functions

5. *Owner_of ():* This function returns the owner of a token's address. Since ERC-721 tokens are unique and non-fungible, they are digitally represented on the blockchain by an ID. This ID can be used by other users, applications, and contracts, to specify who owns the token.
6. *Approve ()*: This function grants another user the authority to transfer NFT tokens on behalf of the owner.
7. *and take ownership():* This function can be invoked by an external party to withdraw tokens from another user's account. Moreover, this function can be used when a user has been approved to own a specified amount of tokens and wants to withdraw those tokens from the balance of another user.
8. *Transfer():* This function enables the token owner to transfer it to another user.
9. *TokenOfOwnerByIndex ():* it is an optional function used to retrieve tokens IDs. Since each owner may own multiple NFTs concurrently, and each NFT is identified by a unique ID, this can become difficult to keep track of IDs over time. Therefore, the smart contract saves these IDs in an array, and the *TokenOfOwnerByIndex()* function enables users to retrieve them.

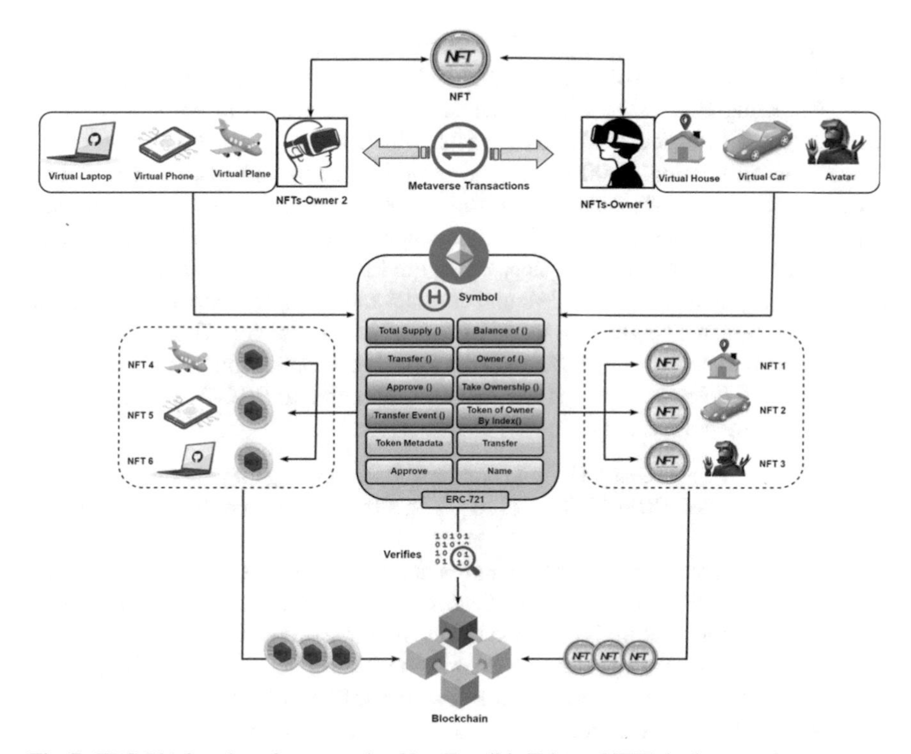

Fig. 7 ERC-721 functions for processing Non-Fungible Tokens (NFTs) in the metaverse

C. Metadata and Events

10. *TokenMetadata:* it is an optional attribute that provides an interface for discovering the metadata associated with a specific NFT token.
11. *Transfer:* This event occurs when the token's ownership changes from one user to another.
12. *Approve:* This event is triggered whenever a user grants another user ownership of the token, i.e., whenever the approve function is executed.

Figure 7 depicts the ERC-721 methodology for processing NFTs tokens in the metaverse.

4.3 ERC-1155 Token Standard

ERC-1155 is an integration between the two previous protocols, ERC-20, and ERC-721. Witek Radomski is the owner of the idea of ERC-1155 as an all-inclusive token standard for the Ethereum smart contracts for processing both fungible and non-fungible tokens [22, 23]. Before ERC-1155, it was not allowed to process fungible and non-fungible tokens with a unique smart contract. This means that if a user wanted to transfer, say, $1000 (i.e. fungible token using ERC-20) and Crypto-Kitties (i.e. NFT using ERC-721) at the same time, he would need to execute multiple transactions with different smart contracts, which was expensive and inefficient. In addition, in gaming, 1000 items (e.g. weapons, armor, shields, devices, coins, badges, castles, etc.)—that players can collect, exchange, and trade with one another will require 1000 smart contracts for processing and verifying these items, which is very redundant and inefficient usage. Using the ERC-1155 token standard, various items can be processed by a single smart contract and multiple numbers of items can be encapsulated in a single transaction to one or more recipients. This means that if a user wanted to send a shield to another, a sword to a friend, and 200 gold coins to both, he could do so in only one transaction. Compared to ERC-20 and ERC-721, ERC-1155 protocol has the following characteristics:

D. ERC-1155 can process an infinite number of tokens using a single smart contract, in contrast, ERC-20 and ERC-721 require a different smart contract for each type of token.
E. ERC-1155 processes not only fungible and non-fungible tokens but also semi-fungible tokens, such as concert tickets. They are interchangeable and can be sold for money before issuing the ticket (fungible). However, after issuing the ticket, they lose their pre-show value and become collectibles (non-fungible).
F. ERC-1155 ensures safe transactions using a safe transfer function that allows tokens to be re-corrected if they are sent to a false address.
G. ERC-1155 verifies and signs fewer transactions for many items as it removes the requirement to "approve" individual token contracts separately.

Gaming is the common application to realize how ERC-1155 works. Its functionality is described through the following functions and attributes:

- *Batch Transfer():* Transfer multiple items in a single transaction.
- *Batch Balance():* Get the balances of multiple assets in a single call.
- *Batch Approval():* verifies and approves all tokens to an address.
- *Hooks():* Receive tokens hook. When the smart contract receives the hook value, this means that the contract accepts the token transferee and processes it as ERC-1155-based tokens.
- *NFT Support:* If the supplied attribute is 1, manipulate this token as NFT.
- *Safe Transfer Rules:* refers to a set of rules for secure token transfer.

Figure 8 depicts the ERC-1155 methodology for processing fungible and NFTs tokens in the metaverse.

5 Blockchain Applications in Metaverse

While metaverse technology is still a relatively new concept and didn't completed as a fully virtual environment, Blockchain is utilized in many applications:

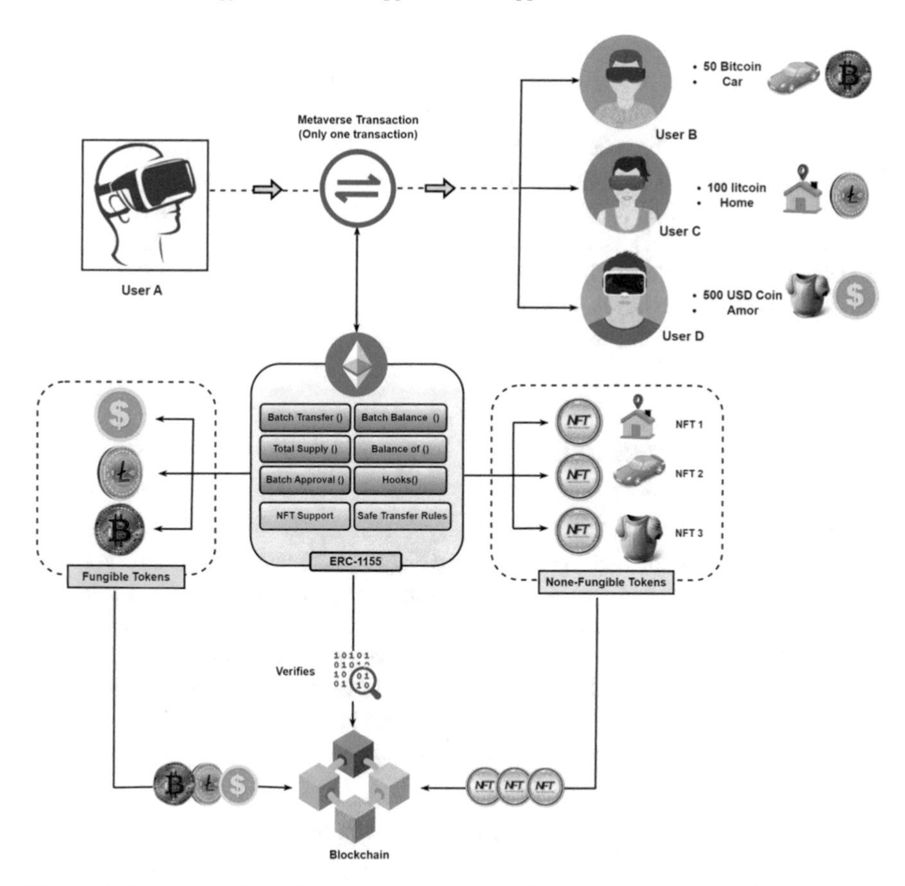

Fig. 8 ERC-1155 functions for processing fungible and non-fungible tokens (NFTs) in the metaverse

5.1 Decentraland (MANA)

Decentraland [24] is the first fully P2P virtual environment and is the largest metaverse cryptocurrency platform built on Ethereum. Moreover, the decentralized autonomous organization (DAO) is the developer and owns the so-called key smart contracts (called a LAND contract) that control and manage the Decentraland metaverse. This type of smart contract allows metaverse users to buy and sell virtual lands on the Decentraland network. MANA is the metaverse's cryptocurrency of Decentraland, which can be used to buy LAND, avatars, wearable objects, names, and other assets in the Decentraland marketplace. The work methodology of Decentraland is based on a program, which is developed to monitor LAND tokens, which addresses real estate properties in the metaverse. This program utilizes the Ethereum blockchain to verify the ownership of this virtual land. Metaverse users must keep their MANA coins in an Ethereum wallet to enable them to create transactions within

Decentraland. The Decentraland system is realized as having two main layers. The top layer keeps track of the virtual land part through a Blockchain-based ledger. Each LAND portion contains a virtual world address, ID, a landowner, and a link to a descriptive file. All these describe describes a specific virtual LAND in the Decentraland environment. The second layer is the content layer. It is responsible for what happens in a user's LAND portion in the metaverse through three important files: (1) content file, which is used to store the digital representation of a LAND portion in Decentraland. (2) Script file, which is used to define how and where the relevant content appears and behaves. (3) Interaction file, which is used to store End-to-end transaction details in Decentraland.

5.2 The Sandbox (SAND)

Sandbox [25] is the second most common cryptocurrency in the metaverse. The original currency used in the metaverse is the SAND virtual token. Using this cryptocurrency, metaverse users can own, develop, and earn their gaming experiences using the SAND Token on the Ethereum Blockchain. Sandbox's users can create virtual assets in the form of non-fungible tokens (NFTs), upload them to the marketplace, and use them in-game transactions as mentioned in Decentraland. The Sandbox functionality is based on two software subsystems, voxEdit and Game Maker. Using both of those subsystems, users can generate their own NFTs, including avatars, virtual goods, and other assets used in the Sandbox game:

1. *VoxEdit* is a software system used by sandbox players to generate and animate their voxel-based NFTs. Voxels are square 3D pixels that are similar to lego blocks. They can be rapidly edited on VoxEdit to form multiple 3D shapes. For instance, players can create avatar-oriented equipment like weapons, and shields or design animals, goods, greenery, and game tools to use in The Sandbox. These 3D objects can then be exported and traded on the Sandbox Marketplace as NFTs tokens.
2. *Game Maker* is a simple software subsystem that allows sandbox users to design and test their unique 3D games within the Sandbox metaverse. players can form and organize different items and objects, including the NFTs created with VoxEdit, in a 3D environment called LAND. For example, they can organize characters and buildings, and edit terrain. They can also share their designed 3D models with the Sandbox marketplace.

5.3 Enjin Coin (ENJ)

Enjin [26] is a platform that allows game creators to issue fungible and non-fungible assets as Ethereum tokens in the game easily. It is based on the ERC-20 token standard that is intended to be utilized as in-game money for buying game objects as well as adding value to in-game assets. In addition, ENJ, the used cryptocurrency token, may also be used as a means of exchange for selling and buying in-game products based on the Ethereum blockchain. Furthermore, ENJ allows players to register unique game elements such as characters or accessories in the form of marketplace tokens, which can then be liquidated for ENJ when needed. using the Enjin digital wallet, players can thus access different partner gaming platforms, keep track of the various tokens or digital assets and sell these assets for ENJ at their convenience. The Enjin platform provides a list of characteristics through its public API and software development kits (SDKs). These features involve: (1) the possibility of virtual goods creation and management, (2) developing multiple software development kits (SDKs), (3) using a digital wallet for cryptocurrency exchange, (4) Enjin Beam, a digital asset distribution system, allowing developers to send thousands of NFTs/FTs to users with a simple QR code. (5) processing REC-20, and ERC-1155 tokens.

5.4 Bloktopia

Bloktopia's [27] metaverse project aims to create a different platform where users may meet new people, study, create a business, and experiment with a variety of activities over a 21-layer structure. Bloktopia also aims to design new methods for creating content and revenue for those who want to live in the metaverse. The native cryptocurrency in Bloktopia is named BLOK. It is an NFT token with which users can purchase things, and access exclusive events within the Bloktopia metaverse. BLOK is used also to buy customized avatars or hire virtual real estate in the Bloktopia environment. For example, it is used for updating real estate space or purchasing 3D objects from the Bloktopia marketplace. What is distinguishing Bloktopia is that it is based on the Polygon (MATIC) network, which is both a cryptocurrency and blockchain scaling platform that aim to create a multi-chain blockchain ecosystem compatible with Ethereum. besides Bloktopia use cases such as access to virtual educational tools about crypto, and virtual events, Bloktopia users can unlock multiple streams of passive and active income, as well as stakes. One of the monetization opportunities is an option to purchase real estate blocks in the form of NFTs, which can later be hired to advertise parties or used to host Bloktopian events.

6 Challenges and Open Problems

Metaverse is still in its early stages and a lot of challenges and open questions are still required great effort to answer, however, Blockchain technology can grant the answer keys to these challenges and questions, especially when integrated with the key enabling technologies in the metaverse. This section highlights these challenges and clarifies how Blockchain can be a magic solution to these challenges [6].

6.1 Tracking AI-Powered Avatars in Metaverse

AI is one of the most important supporting technologies which can be used in many applications such as the space industry [28]. Currently, AI can be used for developing multiple machines and deep learning applications metaverse. Using AI, avatars-3d shapes can be predicted from images, and avatars-static and dynamic features can be utilized for developing clustering, classification, and prediction techniques in the metaverse. AI will be also responsible for manipulating the feelings, facial expressions, and behaviors of avatars. However, making the metaverse so entertaining and authentic at the same time is an important challenge. Developing malicious AI bots for winning games, or stealing 3d assets across the metaverse is an additional challenge. Users have no means to distinguish whether they are interacting with a real person or a fake computer-created avatar.

The encryption capabilities of Blockchain will protect the highly sensitive data that AI-driven systems require, store, and use. Indeed, public Blockchain is secure and immutable, however, this will not prevent an attacker to develop fake AI avatars which can do transactions doesn't follow Blockchain protocols and consensus due to the open nature of the metaverse. Hence, tracking and verifying the identity of AI-based avatars is a major problem in Metaverse.

6.2 Sensing Metaverse Data Using IoT Networks

In the Metaverse, a virtual platform will require collecting data from multiple Internet of Things (IoT) devices to guarantee that it performs efficiently in various metaverse applications such as education, medicine, agriculture, and other metaverse environments. Using special types of sensors, IoT devices equipped with those sensors will connect to the metaverse and will be able to navigate physically and virtually in many virtual environments. However, the capacity of IoT devices to process metaverse operations will be critical to the potential of sensing metaverse data using IoT networks. With so many connected devices, IoT security, and unstructured data storage and processing are undoubtedly a great challenge in the metaverse. In addition, the huge number of connected heterogeneous IoT devices in various virtual domains in the metaverse makes the centralized techniques for manipulating metaverse data and operations not an effective option at all.

Using Blockchain, Metaverse users and applications will be able to access and process IoT data depending on decentralized management and control techniques. Blockchain will enable IoT devices to produce tamper-resistant records of shared transactions in virtual worlds due to Blockchain transparency characteristics. Thanks to Blockchain authenticity, metaverse users will be able to do various transactions in robust and confidential manners, where each transaction will be recorded and authenticated according to the defined Blockchain consensus or protocols. Blockchain will enable IoT nodes to store and share real data securely across a variety of virtual environments within the metaverse. However, this type of transaction will require a significant amount of processing power to keep them running in the metaverse efficiently. Therefore, *Blockchain processing power* will be an open problem and an important challenge in the metaverse. Moreover, the property of distributed transaction ledger of Blockchain can be negatively exploited by some users for developing unlawful or malicious Smart contracts. Therefore, *Blockchain anonymity* is another major challenge while sensing IoT data in the metaverse.

6.3 Synchronization Between Actual and Virtual Worlds Using Digital Twins

A Digital twin [29] is a digital representation of all assets, systems, resources, functions, and operations in the actual world and transforms them into digital counterparts to work for a particular purpose in the virtual world. The metaverse applications will not be able to work efficiently unless a synchronous connection between the physical and virtual worlds. Digital twins will be a base technology to understand how the metaverse system will evolve and will aid in the prediction of the virtual life future. The challenge is how to digitally store those digital twins. How to verify their transactions across various domains in the metaverse? How to ensure the accuracy and quality of digital twin data? Metaverse domains are constantly changing, therefore, digital twins in the metaverse have to detect and respond to these changes. In addition, securing digital twin data against malware and fake digital twins-agents is a further important challenge.

The transparency feature and historical data archiving of Blockchain can ensure tamper proof of digital twin data. Linking blocks using hashing and encryption techniques will enable digital twins to resist metaverse attacks and securely share data between various digital twins. Moreover, by integrating blockchain with AI, it will be possible to track the sensed data from physical worlds and produce high-quality digital twins in the metaverse. Storing digital twin actions in the metaverse as a set of transactions on the Blockchain will make these transactions immutable and impossible to alter. However, *privacy, standardization, and scalabilit*ies are all open issues that must be addressed for blockchain to be successfully mapped and synchronous physical systems into the corresponding digital twins in the metaverse.

6.4 Big Data Storage in Metaverse

Data in the metaverse will be more diverse and sent/receive in greater volumes with high velocity than data in the physical world. Although data storage technologies are different and have advanced, the size of data has progressively doubled and will continue to rise in the metaverse. The arrangement and compromising of the size and speed of created data in the metaverse is a complex challenge in virtual worlds.

How Blockchain can mitigate this challenge? Blockchain technology will help in storing data from authenticated and trusted data sources; hence, improper data will be erased from the metaverse. Therefore, *data cleaning* will be an important characteristic of Blockchain for storing big data volumes in the metaverse. However, storing data in limited sized-blocks may not be prober to the increasing volumes of data in the metaverse. Therefore, storing large volumes of metaverse data patterns is still a big open issue.

6.5 Manipulating Holographic Telepresence in the Metaverse

VR, AR, and ER are the main technologies that will be used for designing holographic telepresence in the Metaverse. Storing, and verifying all transactions of holographic models in real-time are important challenges that cannot be solved by traditional and known storage and verification techniques.

Designing Blockchain as distributed ledger would help verify holographic models and other XR applications data and trace the source of erroneous data using smart contracts. This will help build more accurate VR/AR/XR applications in the metaverse. In addition, the transparency, security, and immutability of Blockchain will support digital twins, IoT, and XR applications to manipulate metaverse data in high quality, secured acquisition, and authentic manners. However, holographic data storage and interoperability are still open problems in the metaverse.

7 Conclusion

The main goal of the current study was to determine the contributions of Blockchain technology as a key technology in the metaverse. This chapter has identified the main characteristics of Blockchain such as security, trust and traceability, smart contract, decentrality, and immutability, and how they can be implemented in the metaverse. The second major aspect was clarifying the difference between fungible and non-fungible (NFTs) tokens and how Ethereum Request for Comments (ERC) protocols, ERC-20, ERC-721, and ERC-1155 can be utilized for processing fungible and non-fungible (NFTs) tokens. In addition, the chapter introduced some examples of real Blockchain applications in the metaverse. Finally, the chapter has ended with some opportunities, challenges, and open problems of utilizing Blockchain in the metaverse.

References

1. Mystakidis, S.: Metaverse. Encyclopedia. **2**(1), 486–497 (2022)
2. Anter, A.M., Moemen, Y.S., Darwish, A., Hassanien, A.E.: Multi-target QSAR modelling of chemo-genomic data analysis based on extreme learning machine. J. Knowl.-Based Syst., Elsevier, **188**, 104977 (2020). https://doi.org/10.1016/j.knosys.2019.104977
3. Jeon, H.J., Youn, H.C., Ko, S.M., Kim, T.H.: Blockchain and AI Meet in the metaverse. Adv. Converge. Blockchain Artif. Intell. **12**, 73 (2022)
4. Ynag, Q., Zhao, Y., Huang, H., Zheng, Z.: Fusing blockchain and AI with metaverse: a survey. arXiv preprint arXiv:2201.03201. 2022 Jan 10
5. By Mike Isaac, Meta spent $10 billion on the metaverse in 2021, dragging down profit. https://www.nytimes.com/2022/02/02/technology/meta-facebook-earnings-metaverse.html#:~:text=Investing%20%2410%20billion%20in%20the,to%20buy%20Instagram%20in%202012. (accessed 12/5/2022)
6. Gadekallu, T.R., Huynh-The, T., Wang, W., Yenduri, G., Ranaweera, P., Pham, Q.V., da Costa, D.B., Liyanage, M.: Blockchain for the metaverse: a review. arXiv preprint arXiv:2203.09738. 2022 Mar 18
7. Crowell, B.: Blockchain-based metaverse platforms: augmented analytics tools, interconnected decision-making processes, and computer vision algorithms. Linguis. Philos. Invest. **21**, 121–136 (2022)
8. Bai, Y., Lei, H., Li, S., Gao, H., Li, J., Li, L.: Decentralized and self-sovereign identity in the era of blockchain: a survey. In: 2022 IEEE International Conference on Blockchain (Blockchain) 2022 Aug 22. IEEE, pp. 500–507 (2022)
9. Acronise website, A blockchain-powered data notarization and e-signature solution for service providers, [online], https://www.acronis.com/en-eu/products/cloud/notary/ (accessed 25/8/2023)
10. Eid, H.F., Darwish, A., Hassanien, A.E., Kim, T.: Intelligent hybrid anomaly network intrusion detection system. In: Communication and Networking. FGCN 2011. Communications in Computer and Information Science, vol. 265. FGCN 2011, Part I, CCIS 265, pp. 209–218 Springer, Berlin, Heidelberg (2011)
11. Landerreche, E., Stevens, M.: On immutability of blockchains. In: Proceedings of 1st ERCIM Blockchain Workshop 2018. European Society for Socially Embedded Technologies (EUSSET) (2018)
12. Christodoulou, K., Katelaris, L., Themistocleous, M., Christodoulou, P., Iosif, E.: NFTs and the metaverse revolution: research perspectives and open challenges. Blockchains and the Token Economy 139–178 (2022)
13. Schinckus, C.: Proof-of-work based blockchain technology and anthropocene: an undermined situation? Renew. Sustain. Energy Rev. **1**(152), 111682 (2021)
14. Saleh, F.: Blockchain without waste: proof-of-stake. The Rev. Financial Stud. **34**(3), 1156–1190 (2021)
15. Saad, S.M., Radzi, R.Z.: Comparative review of the blockchain consensus algorithm between proof of stake (pos) and delegated proof of stake (dpos). Int. J. Innov. Comput. **10**(2) (2020)
16. Mingxiao, D., Xiaofeng, M., Zhe, Z., Xiangwei, W., Qijun, C.: A review on consensus algorithm of blockchain. In: 2017 IEEE International Conference on Systems, Man, and Cybernetics (SMC) 2017 Oct 5. IEEE, pp. 2567–2572 (2017)
17. Andrey, A., Petr, C.: Review of existing consensus algorithms blockchain. In: 2019 International Conference Quality Management, Transport, and Information Security, Information Technologies"(IT&QM&IS) 2019 Sep 23. IEEE, pp. 124–127 (2019)
18. Ali, M., Bagui, S.: Introduction to NFTs: the future of digital collectibles. Int. J. Adv. Comput. Sci. Appl. **12**(10), 50–56 (2021)
19. Nathan Reiff, What Are ERC-20 Tokens on the Ethereum Network? https://www.investopedia.com/news/what-erc20-and-what-does-it-mean-ethereum/ (access 10/11/2022)

20. Arora, A., Kumar, S.: Smart contracts and NFTs: non-fungible tokens as a core component of blockchain to be used as collectibles. In: Cyber Security and Digital Forensics 2022. Springer, Singapore, pp. 401–422
21. Takyar, A.: How to create ERC-721 Token? https://www.leewayhertz.com/create-erc-721-token/ (accessed 15/11/2022)
22. MoreReese, What is ERC-1155? Ethereum's Flexible Token Standard. https://decrypt.co/resources/what-is-erc-1155-ethereums-flexible-token-standard (accessed 23/11/2022)
23. Muthe, K.B., Sharma, K., Sri, K.E.: Blockchain-based decentralized computing and NFT infrastructure for game networks. In: 2020 Second International Conference on Blockchain Computing and Applications (BCCA) 2020 Nov 2. IEEE, pp. 73–77 (2020)
24. Goanta, C.: Selling LAND in Decentraland: the regime of non-fungible tokens on the Ethereum blockchain under the digital content directive. In: Disruptive Technology, Legal Innovation, and the Future of Real Estate 2020. Springer, Cham, pp. 139–154
25. Wang, X., Chen, Q., Li, Z.: A 3D reconstruction method for augmented reality sandbox based on depth sensor. In: 2021 IEEE 2nd International Conference on Information Technology, Big Data and Artificial Intelligence (ICIBA) 2021 Dec 17. IEEE, vol. 2, pp. 844–849 (2021)
26. SEİFODDİNİ J. A multi-criteria approach to rating Metaverse games. J. Metaverse. **2**(2), 42–55 (2022)
27. Türk, T.: The concept of metaverse, its future and its relationship with spatial information. Adv. Geomatics **2**(1), 17–22 (2022)
28. Abdelghafar, S., Darwish, A., Hassanien, A.E.: Intelligent health monitoring systems for space missions based on data mining techniques. In: Hassanien, A., Darwish, A., El-Askary, H. (eds.) Machine Learning and Data Mining in Aerospace Technology. Studies in Computational Intelligence, vol. 836. Springer, Cham (2020). https://doi.org/10.1007/978-3-030-20212-5_4
29. Far, S.B., Rad, A.I.: Applying digital twins in metaverse: user interface, security and privacy challenges. J. Metaverse. **2**(1), 8–16 (2022)

The Threat of the Digital Human in the Metaverse: Security and Privacy

Mona M. Soliman, Ashraf Darwish, and Aboul Ella Hassanien

Abstract A metaverse is a collection of shared, multi-user, persistent, 3D virtual worlds that are connected to the real world and combined to form a single, eternal virtual reality. The metaverse is transitioning from science fiction to an imminent reality thanks to recent developments in new technologies including blockchain, augmented reality, digital twins, extended reality, and augmented reality. Users enter the metaverse with avatars that they can use to interact with other users as well as the objects, programmes, services, and organisations that make up the metaverse. However, this new mode of communication raises a lot of privacy and security issues. Threats related to the technologies that will be used to realise the metaverse will grow in quantity and quality, and some equally unheard-of threats will also emerge. The privacy and security of users of the metaverse, which we discuss in this chapter, are particularly vulnerable to such threats. MSPs (Metaverse Service Providers) should consider these issues. With ternary-world interactions, we first study an unique distributed metaverse architecture and its important features. The privacy threats to the metaverse are then discussed, along with security countermeasures for these threats. We also examine the primary security concerns associated with the development of the metaverse and data exchange within it.

1 Introduction

The phrase "metaverse," which combines the words "meta" and "universe," refers to a shared virtual world that is supported by a number of emerging technologies, including fifth-generation networks, beyond virtual reality, Augmented Reality,

Mona M. Soliman, Ashraf Darwish, Aboul Ella Hassanien: Scientific Research Group in Egypt (SRGE). http://www.egyptscience.net

M. M. Soliman (✉) · A. E. Hassanien
Faculty of Computers and Artificial Intelligence, Cairo University, Cairo, Egypt
e-mail: mona.solyman@fci-cu.edu.eg

A. Darwish
Faculty of Science, Helwan University, Cairo, Egypt

A. E. Hassanien et al. (eds.), *The Future of Metaverse in the Virtual Era and Physical World*, Studies in Big Data 123, https://doi.org/10.1007/978-3-031-29132-6_14

Extended Reality, digital twins, block chain and artificial intelligence (AI) [1]. A form of embodied internet that you are inside of rather than just looking at, the Metaverse is a virtual environment where you may be present with others in digital areas, according to Mark Zuckerberg, CEO of Facebook. In the next few of years, metaverse will replace mobile internet. The majority of young people currently interact with their phones for 7–10 h every day; they already live on Facebook, YouTube, Instagram, Tik-tok, etc.; and soon they will be able to appear inside of these platforms as avatars, much like some people have already gone to a Travis Scott concert in the video game Fortnite [2]. Augmented reality (AR) will enable seamless interaction between the virtual and real worlds, virtual reality (VR) will be used to build immersive 3D spaces. Digital twins will enable the transfer, visualization, and sharing of physical items into the metaverse along similar lines. While some sensors, such as those built into the next generation of smart gadgets, will feed more real-world data into the metaverse, wearable sensors will enable avatars in the virtual worlds to emulate real-world movements.

The metaverse may inherit the weaknesses and inherent defects of the various cutting-edge technology that it incorporates and the systems that are based on them. Emerging technology incidents have included the theft of virtual currencies, the appropriation of wearable technology or cloud storage, and the misuse of AI to create false news. In virtual worlds, the consequences of current risks can be exacerbated and made more severe, while new threats that do not exist in real or cyberspace can emerge, like virtual stalking and virtual spying [3]. In particular, the metaverse's use of personal data to create a digital replica of the real world can be more granular and pervasive than ever before, which creates new opportunities for crimes involving private big data [4]. Users will surely employ wearable AR/VR devices with built-in sensors to collect a wide range of data, such as brain wave patterns, facial expressions, eye movements, hand movements, speech patterns, biometric traits, and the environment [5]. Many sensitive user behaviors will be processed in real-time by various hardware vendors. Wearables and other internet-connected devices enable for user tracking and data collection. Numerous interpretations are possible for this information. These gadgets have the ability to gather a variety of data, including communications, user behaviour (such as habits and decisions), and personal information (such as physical, cultural, and economic data) (e.g., metadata related to personal communications). Last but not least, hackers can leverage compromised devices and system flaws as entry points into real-world equipment like home appliances to endanger personal safety and even endanger key infrastructures like water supply systems, high-speed train systems, and power grid systems [6].

Most of the time, like with smart homes, we aren't even aware that such pervasive and ongoing recordings are taking place, which puts our privacy at risk in unexpected ways [7]. Zuckerberg acknowledges the security issues that the metaverse inevitably raises. In the metaverse, various assets need to be protected, including user profiles with behaviour data, authentication credentials, and sensitive biometric data. Data for digital goods and assets, such as non-fungible tokens (NFTs), and cryptocurrency are next, followed by systems for delivering services and applications [8]. The administration of these enormous data streams, ubiquitous user profiling practises, unfair results of AI algorithms, and the safety of physical infrastructures and human bodies are only a few examples of the many security flaws and privacy invasions that could occur in the metaverse [3].

In this Chapter we will elaborate on the privacy and security risks that individuals can face when using the metaverse. We start in Sect. 2 introducing a brief introduction about the main characteristic features of metaverse and introducing the enabling technology that can be used to build metaverse world. Then, we explore the privacy issues related to metaverse by analyzing both threats and Security Countermeasures for these threats in Sect. 3. Section 4 introduces the security issues related to metaverse data. Finally, Sect. 5 provides conclusions about the security and privacy threats, as well as the critical challenges in security defenses and privacy preservation.

2 Metaverse: Concepts and Characteristics

Internet users are merely content consumers in web 1.0, when websites offer the material. Users are both content producers and consumers in web 2.0 (also known as mobile Internet), and websites function as platforms for service delivery. The web 3.0 paradigm known as the metaverse is widely acknowledged.

Users who are represented by digital avatars can quickly switch between several virtual worlds (i.e., sub-metaverses) in the metaverse to live a digital existence and conduct business. Physical infrastructures and many enabling technologies enable this. To enrich our experiences with extended reality and improve many aspects of the digital world, technological evolution is crucial. A highly developed 3D environment that can provide individuals with an immersive digital experience was introduced after many technologies combined to replicate the metaverse's fundamental idea. We begin by looking at the fundamental ideas behind the metaverse and its unique features. We look into the administration and control of information in the metaverse. A quick summary of the primary features of the metaverse, the technology that was utilized to develop it, and the fundamental components of the global infrastructure are shown in Fig. 1 and discussed in more details in the following sections.

2.1 *Basic Concepts*

The metaverse enables users to connect with other user-controlled elements and manipulate digital elements in a shared virtual environment, simulating real-world social interactions. Users can, for instance, operate their digital avatars for everyday communication, collaboration, sharing, and gaming in the metaverse zone by using XR interaction devices. Users can swiftly build, navigate through various metaverse zones, and produce digital assets in the meantime. The digital world can be made up of a number of interconnected distributed virtual worlds (i.e., sub-metaverses), according to ISO/IEC 23,005 and IEEE 2888 standards [11, 12]. Each

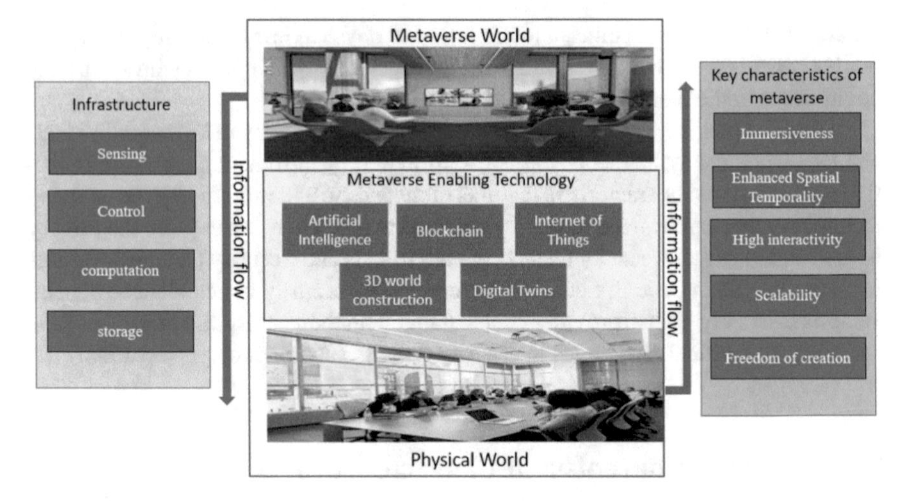

Fig. 1 Enabling Technology, key characteristics, and infrastructure of the metaverse

sub-metaverse can provide users portrayed as digital avatars with particular types of virtual goods/services (e.g., gaming, social dating, online museums, and online concerts) and virtual environments (e.g., game scenes and virtual cities). In the metaverse, users' digital avatars are referred to as avatars. In numerous metaverse applications, users can create a variety of avatars. These avatars can take the form of humans, animals, fantastical creatures, etc. In the metaverse, "virtual environments" refers to the simulated actual or imagined surroundings (composed of 3D digital objects and their properties). In addition, the metaverse's virtual worlds can have unique spatiotemporal dimensions (such as those in the past or the future) that allow users to experience a different way of life. The term "virtual goods" refers to the tradable items created by virtual service providers (VSPs) or users in the metaverse, such as skins, digital artwork, and land parcels. The metaverse offers a wide range of virtual services, such as a digital market, digital currency, digital regulation, social services, etc.

2.2 Key Characteristic of Metaverse

The metaverse exhibits the fundamental traits listed below [9], creating a profoundly connected and interactive immersive spatial platform surrounding the real world, the virtual world, and people:

- Immersiveness: Users can feel mentally and emotionally immersed in virtual space since it is sufficiently realistic. It is often referred to as immersive realism. The virtual digital world in the metaverse needs to be sufficiently realistic, such as through the reconstruction of digital avatars, in order to achieve the greatest level of integration with the physical world and allow users to be visually immersed. Realistically speaking, people connect with their surroundings through their senses and bodies. The structure of sensory perception (such as sight, sound, touch, temperature, and balance) and expressiveness (such as movements) can be used to approach immersive realism [10]. As a result, users can synchronise, transmit, and interact with these perceptual signals between several digital avatars to achieve immersion from a physiological perspective.

- Enhanced Spatial Temporality: Real-time human interactions across space and time are constrained by spatial distance and are not possible in the physical world due to these limitations. The metaverse, in contrast, is a hyperspace setting parallel to the real world that can transcend time and space [10]. Users can freely travel between numerous worlds with diverse spatiotemporal dimensions, for instance, to experience a different way of life with seamless scene alteration. Even faster switching between live situations, places, and timelines is possible for users, something that is currently not possible in the real world.
- High interactivity: Using the digital avatar, users can move about and communicate in various metaverse zones. Additionally, they can coordinate, communicate, and provide feedback on the characteristics and conditions of the environment in both the digital and actual worlds. The user-created digital assets can converse with one another. Additionally, separate virtual and real worlds have unique interaction tasks and themes in their respective metaverse zones, and these duties and the states of the metaverse elements are always changing. Users can move between virtual worlds without the immersive experience being interrupted, and different systems can use the same digital assets to create or reconstruct virtual worlds [10]. As a result, coordinating communication and providing feedback is a key aspect of metaverse interaction.
- Scalability: The term "scalability" refers to the ability of the metaverse to remain effective despite changes in scene complexity, the number of concurrent users/avatars, and the type, scope, and range of user/avatar interactions [10].
- Freedom of creation: Users are given stronger independence and creative space by the metaverse, which also offers a secure and independent user environment. They can produce digital assets in the virtual world and produce money by moving around in various metaverse zones. The metaverse can also constantly pique consumers' interest in open innovation and the development of digital content.

2.3 *Enabling Technology in Metaverse*

To enrich our experiences with extended reality and improve many aspects of the digital world, technological evolution is crucial. A highly developed 3D environment that can provide individuals with an immersive digital experience was introduced after many technologies combined to replicate the metaverse's fundamental idea. Virtual reality (VR) and augmented reality (AR) technologies are frequently used to characterize the metaverse, although metaverse development goes beyond these two fundamental technologies. Blockchain is a key component of the many technologies driving the metaverse. Blockchain technology has the ability to make metaverse projects compatible with Web 3, which is the next phase of the internet, by offering a decentralized infrastructure to metaverse and encouraging the development of compelling use cases for its ecosystem.

Another significant factor driving the growth of the metaverse is artificial intelligence. Businesses can replace human activities with automated, computer-controlled ones by deploying AI. Fast computing, identity verification via facial recognition, analytics, and scaling better techniques are the most often used applications of AI for enterprises. Regarding the employment of AI in the metaverse, it enables the creation of 2D and 3D based avatars based on the unique traits of users.

IoT is a technology that uses sensors and other devices to act as a conduit between the physical world and the internet. IoT strengthens the link between the metaverse and physical things or gadgets in the physical world. These gadgets can effortlessly transmit and receive information once they are connected to the metaverse, improving how accurately the physical world is replicated there. IoT also gives the metaverse access to real-time information gathered from the outside world. By making use of such information, the metaverse can improve the veracity of the events taking place in its virtual world and make them pertinent to the circumstances or environment in the actual world.

A digital twin (DT), which is another enabling technology that is important for creating the metaverse. It is a virtual object that is linked to a physical object, such as a robot or piece of machinery. IoT systems and software that are used to build a digital representation of the physical asset assist this mapping of the digital world to real-world assets. To reconstruct and depict a high-fidelity digital reality, digital twin technology creates digital clones of diverse elements and multidimensional situations in the actual world. In the digital realm, it can also predict and optimize autonomously before returning the choice or warning outcomes to the consumers and the real world. As a result, we can foresee, regulate, trace, and do other tasks using digital twin technology, hence lowering the likelihood of mishaps and hazards in the physical world.

2.4 Information Flow Among Metaverse World

Wang [3] describes the metaverse as a synthetic world made up of user-controlled avatars, digital objects, virtual environments, and other computer-generated elements where people (represented by avatars) can interact, collaborate, and socialise with one another using any smart device and their virtual identities. The blockchain, digital twin, AI, and blockchain-based metaverse engine [13] creates, updates, and maintains the virtual world using large data from the actual world as inputs. Users located in physical environments can immersively control their digital avatars in the metaverse via their senses and bodies for various group and social activities like car racing, dating, and virtual item trading with the help of XR and HCI (especially brain-computer interaction (BCI)) techniques In the metaverse, the virtual economy can be developed as an organic byproduct of such avatar digital production activities. To improve the metaverse ecosystem, AI algorithms create customised avatars and content and render the metaverse on a big scale. Additionally, using digital twin technology, the knowledge obtained through AI-based big data analytics may be used to simulate, digitalize, and mirror the real world to create lifelike virtual environments for people to experience. As a last step toward creating the economic system and value system in the metaverse, the manufactured digital twins as well as native contents made by avatars can be openly managed, uniquely tokenized, and monetized using blockchain technology. There are a lot of information that flow among the metaverse world.

- In-World Information Flow: Human society or the world is connected by social networks and is built on the mutual contacts and common activities of individuals. Through the use of ubiquitous sensors and actuators, IoT-enabled sensing and control infrastructure significantly contributes to the digitalization and transformation of the physical world. Network and computing infrastructures are used to transmit and analyse the IoT big data that is produced.
- In the digital world: The metaverse engine processes and manages the created digital information of the physical and human worlds to facilitate massive metaverse creation/rendering and numerous metaverse services. In addition, users who are represented by avatars can create and share digital works across different sub-metaverses to encourage metaverse creativity.
- Information Flow Across Worlds: The three worlds' primary media are subjective consciousness, the Internet, and the Internet of Things. I Through the use of HCI and XR technologies, humans can interact with real-world items and experience virtual augmented reality (such as holographic telepresence). (ii) The Internet, the biggest computer network in the world, connects the physical and digital worlds. Smart gadgets, such as smartphones, wearable sensors, and VR headgear, allow users to interact with the digital environment for the creation, sharing, and learning of knowledge. (iii) The IoT infrastructure connects the physical and digital worlds by digitalizing them with interconnected smart devices, allowing information to easily move between them [14]. Additionally, the feedback data from the digital world (such as the analysed outcomes of big data and intelligent judgements) can direct the transformation of the physical world (such as the manufacturing process). Threats in virtual worlds can be multiplied and seriously harm actual infrastructures and personal safety because the metaverse combines physical systems, human civilization, and cyber worlds. This creates enormous demands and challenges for governance.

3 Metaverse Data Protection and Privacy

Privacy In the modern Internet, it is stated that if you don't pay for a good or service, then you (or, more specifically, your data) are the thing that is being sold. The best illustration of this kind is social media and social networking sites. These platforms provide free services that involve millions, if not billions, of users, all of whose preferences are well-known to the platforms themselves [15], allowing them to display incredibly precise, microtargeted adverts to consumers. Only because of the platform's ability to accurately profile its users through analysis of their actions and interactions with the platform's content and other users, rather than relying on more concerning tracking capabilities, thanks to the evolution of cookies and, generally speaking, fingerprinting techniques [16], is this successful business model possible. The digital traces we leave behind us, even with today's technology, already reveal a lot about our personalities, preferences, and orientations (e.g., political and sexual). Due to the fact that 20 min of virtual reality (VR) use generates almost two million unique data components, the threat to privacy in Web3 and the metaverse is larger than it is in Web2. These include, among many other things, how you breathe, walk, think, move, and stare. To get information, the algorithms map the user's body language. Consent is almost impossible to get in the metaverse because data collecting is ongoing and involuntary.

How we interact socially, learn, shop, play, and travel is changing as a result of the metaverse. Along with the great changes it will bring, we should be ready for any potential negative outcomes. Additionally, because the metaverse will gather more user data than ever before, the consequences if something goes wrong will likewise be harsher than before. The vulnerability to personal information is among the main issues [17]. For example, the technology behemoths Amazon, Apple, Google (Alphabet), Facebook, and Microsoft have long supported passwordless authentication, which establishes identity using a fingerprint, face recognition, or a PIN. This trend is undoubtedly going to continue in the metaverse, probably with even more biometrics like iris and auditory recognition [18]. In the past, if a user lost their password, the worst-case scenario was that they lost some data and had to create a new one to ensure the security of other data. However, because biometrics are linked to a person continuously, if they were compromised (taken by an impostor), they would be permanently compromised and unable to be revoked, putting the user in serious danger [19]. As shown in Fig. 2 we can consider two types of data flow in the metaverse, data related to the identity of the user, where identity authentication and access control are crucial for massive users/avatars in service provided by the metaverse. The other type of data can be called communication data, which refers to data collected by wearable devices. Such data flow is subject to different types of threads. Threats to Metaverse Authentication and Threats to Metaverse Access Control are two categories into which threats to metaverse data authentication can be divided. For wearable devices data users/avatars may suffer from threats in terms of data tampering, false data injection, low-quality UGC, ownership/provenance tracing, and intellectual property violation in the metaverse.

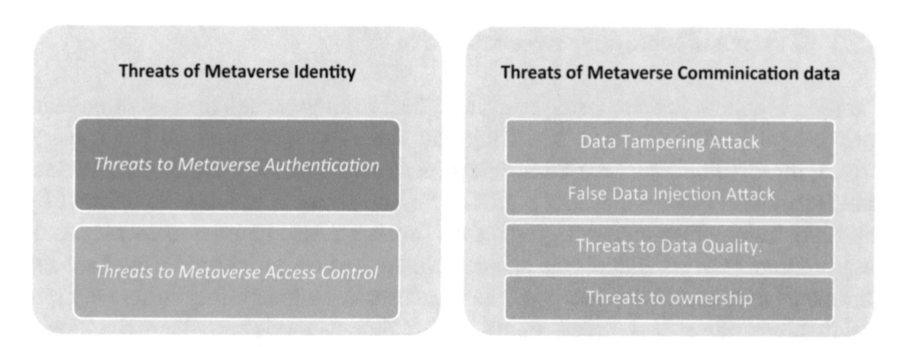

Fig. 2 Threats to metaverse data

3.1 Privacy of Metaverse Identity: Threats and Security Countermeasures

3.1.1 Threats to Metaverse Authentication

When a user or programme tries to access a system, authentication is the procedure used to verify their identity. Because after authentication, the user has access to all the resources. Users' and avatars' identities in the metaverse are susceptible to theft and impersonation, and interoperability problems with authentication can arise across virtual worlds. The consequences of identity theft in the metaverse, which can be more severe than in conventional information systems, can include the loss of a user's avatars, digital possessions, social connections, and even their entire digital life. An attacker can use an impersonation attack to access a service or system in the metaverse by posing as another authorised entity [6]. For instance, hackers can get into the Oculus helmet and use the stolen behavioural and biological data collected by the built-in motion-tracking system to construct digital copies of the user and pose as the victim to aid social engineering assaults. In order to deceive, cheat, and even commit a crime against the victim's friends in the metaverse, the hackers can also make a phoney avatar utilising digital copies of the victim. The identification of avatars for users in the metaverse can be more difficult due to the verification of face features, voice, video footage, and other factors. In addition, enemies can imitate a user's look, voice, and mannerisms to build several AI bots (also known as digital people) that behave exactly like the user's real avatar in the virtual world (such as Roblox) [4]. As a result, more personal data may be needed as proof in order to maintain secure avatar authentication, which could lead to new privacy breach risks. Correct identification in the metaverse will result in the construction of secure settings where all those operations and activities can be carried out.

3.1.2 Threats to Metaverse Access Control

For the generation and rendering of avatars, data from users and their environments must be collected and processed, and this data can be compromised. Different users' private information may be used in violation of laws like the General Data Protection Regulation (GDPR). A user's privacy and preferences can also be inferred by attackers from published processing results (avatars). Also In order to enable seamless tailored services like avatar creation, many service providers throughout the sub-metaverses must have access to real-time user behaviour. Using buffer overflow and manipulating access across control lists, malicious service providers can illegitimately increase their data access permissions. Malicious VSPs may launch assaults to gain access to unauthorised data in order to profit. In addition, it can be difficult to identify exactly what personal information should be shared, with whom, under what circumstances, for what purposes, and when it should be erased because such vast amounts of personal information are produced and communicated in real time.

User data can be purposely disclosed by hackers or accidentally disclosed by service providers during the data-service lifecycle to support user profiling and precision marketing efforts. It is challenging to track the data misuse activities in the large-scale metaverse due to the potential non-interoperability of some sub-metaverses.

3.2 Privacy of Metaverse Communication Data: Threats and Security Countermeasures

The metaverse includes the continuum of reality and virtuality and enables users to transition between them seamlessly thanks to continuous, omnipresent, and universal access to information [23]. Because they enable comfortable user mobility, mobile and wearable devices, such as AR glasses, headsets, and smartphones, are currently the most appealing and extensively used metaverse interfaces. A user's avatar is created using a profile of their facial expressions, eye and hand movements, speech, biological characteristics, and brain wave patterns. The Oculus headset, for instance, has four built-in cameras and motion sensors that may detect our environment and be used by attackers. Sensitive user data is transferred over wired and wireless communication when it is captured by XR devices like headsets. Even if this sensitive data is encrypted, hackers can still intercept it through several means to gain access to the raw data. The position of a user is tracked using differential attacks and sophisticated inference techniques.

3.2.1 Data Tampering Attack

Integrity characteristics keep track of any alterations made to data while it is transmitted across ternary worlds and sub-metaverses. Attackers can alter, delete, and replace metaverse data services at any time to obstruct users', avatars', or physical entities' regular operations [24]. These attackers can conceal their illicit activity in the virtual world by avoiding detection by relevant log files or message-digest results.

3.2.2 False Data Injection Attack

Attackers can trick metaverse systems by injecting fabricated data, such as fraudulent messages and incorrect instructions [26]. By inserting poisoned gradients or opponent training samples (centralised or decentralised) during training, for instance, attackers might produce biassed AI models. The security of physical equipment and even personal safety may be threatened by incorrect feedback or instructions that are returned. For instance, fabricated feedbacks like too much voltage can harm and malfunction wearable XR gadgets. Another illustration is the possibility that a human user can pass away due to Fortnite's (a metaverse game) altered 100-fold magnifications of physical pain upon being shot.

3.2.3 Threats to Data Quality of UGC and Physical Input

Users who produce low quality content to cut expenses may jeopardise the utility of user generated content (UGC), such as data quality. During the training phase of the content recommendation model, they can exchange non-IID unaligned data. Uncalibrated wearable sensors can potentially produce erroneous data that deceives the development of digital twins. For instance, during the collaborative training phase of the content recommendation model in the metaverse, they can contribute unaligned and severe non-IID data, leading to erroneous content suggestion. Another illustration is how imprecise and even incorrect sensory data from uncalibrated wearable sensors can be used to create digital twins in the metaverse and degrade user experience.

3.2.4 Threats to UGC Ownership and Provenance

Unlike the government-supervised asset registration process in the real world. The metaverse is a decentralised, open, and autonomous environment. As a result, it is challenging to establish the ownership and provenance of user-generated content (UGC) created by numerous avatars across all sub-metaverses [28]. Additionally, because UGCs have digital properties that allow for unlimited replication and real-time sharing within a single virtual world or across multiple virtual worlds, it is more difficult for.

3.3 Security Countermeasures to Metaverse Authentication & Access Control

User/avatar interaction and service provisioning in the metaverse are based on safe and effective identity management. Digital identities can generally be divided into three categories:

- Centralized identity: The government-approved digital ID system, such as a Gmail account, has a single supplier under the centralised paradigm and provides confirmation of legal identification.
- Federated identity: In a federated approach, various organisations work together or are accredited by a trust framework or federation body to offer a digital ID that is recognised by the government. In certain circumstances, these identity providers use a fundamental ID system as their authoritative source. They may be public or private businesses.
- Self-sovereign identity (SSI): Digital identities that are entirely under user control are referred to be SI. It enables users to undertake cross-domain actions to enable identity interoperability with their consent, autonomously sharing and associating various personal information (for example, username, educational information, and employment information).

Centralized identity management solutions may be vulnerable to a *single point of failure* (SPoF) hazards and leakage risks in the metaverse. Federated identity systems are semi-centralized, and a small number of organizations or federations are in charge of managing IDs. These organizations or federations could potentially face centralization problems. Future metaverse construction will be dominated by identification systems based on *SQL Server Integration Services* (SSIs). Identity management solutions in the metaverse should, in accordance with [29], adhere to the following design principles: Scalability to large numbers of users or avatars, node damage resistance, and interoperability across many sub-metaverses during authentication are the first two requirements.

Numerous techniques are being suggested to mitigate the privacy problems in the Metaverse [20]. The user has the capacity to create numerous copies of his avatar, allowing him to displace and change his appearance. Attackers will be perplexed as to which of the many clones of an avatar is the real user if there are enough of them scattered around the metaverse. The user-configurable behaviour of the avatars can vary. For instance, when purchasing an item in the metaverse, the user can create a different avatar that purchases a certain set of products, confusing and distracting the attacker, who will not be able to identify the real avatar. An avatar can teleport to generate many copies of themselves that can be used to block tracking. Companies and governments will nevertheless want to follow the user. Smart contracts can be used to specify the conditions of this. To verify a user's identity in the metaverse, methods including body scanning, facial recognition software, DNA identification, and retinal recognition can be utilised [21]. A second technique makes private, temporary replicas of a section of the metaverse (e.g., a park). As a result of the private component this approach has generated, users cannot be overheard by attackers. whether or not the metaverse's primary fabric's generated replica will produce new objects (for example, store items). The metaverse API should then address the merge from the private copy to the main fabric of the metaverse in the event that the private portion uses resources from the main fabric. When the merge is finished, for instance, the things the user purchased in the private department store should be updated in the main shop. The simultaneous use of many private copies of the same area of the metaverse will inevitably present a number of difficulties. To prevent inconsistencies and a deterioration of the user experience, methods that handle the concurrent use of things in the metaverse should be adopted (e.g., the disappearance of items in the main fabric because they are being used in a private copy). Users may also be permitted to make invisible copies of their avatars after creating privacy copies, allowing them to interact in the metaverse unobserved. When the resources of the main fabric are constrained or shared, this solution will face the same problems as the private copies [27].

Since these are private details, it is crucial to protect the privacy and accuracy of biometric information at this stage. By enabling users to sign in using a mix of two or more authentication methods, another security paradigm is presented. The system is more secure since an attacker must defeat more than one security measure to access it. This model's RubikBiom invention may serve as an example [22]. The multi-modal technique enters the virtual world using a variety of verification models. By using gaze-based authentication, the user creates a unique verification that is incomprehensible to others in the same physical space. When this procedure is used in conjunction with a specified schematic image, the highest level of reliability is guaranteed.

Dynamic Data-Masking is another suggestion for safeguarding data in the IT infrastructure and hiding them from online criminals. It enables the replacement of data with peers that resemble the original data in order to conceal the real circumstances of the data. Through the substitution of alternative data directories for user data in the Metaverse universe, it aids enterprises in developing a cutting-edge cybersecurity system. With its Multi-modal Authentication and Dynamic Data-Making modules, Single Connect, Kron's Privileged Access Management (PAM) [25] solution, may assist businesses in ensuring advanced level security in the metaverse. By building a real-time control system for the virtual world and deploying decoys to divert cyberattackers from their intended targets, Single Connect ensures cybersecurity while enabling authentication and masking data owned by metaverse platforms.

4 Metaverse Security: Threats and Solutions

The metaverse is a virtual universe made of digital recreations of the real world and digital works produced by avatars. Digital twins and UGC as well as avatar behaviours (such as conversation and browsing records) will provide certain value

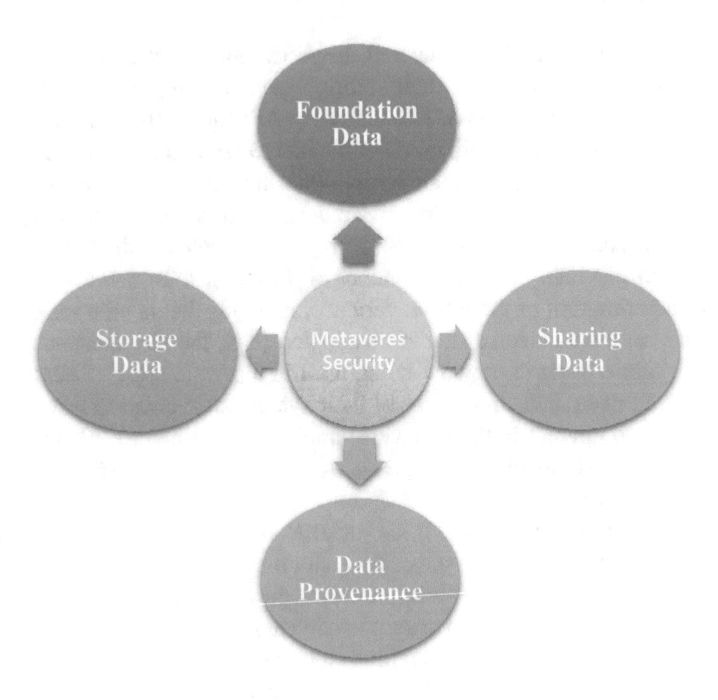

Fig. 3 Security issues for metaverse data

in the metaverse, similar to the value produced by human actions in the actual world [30]. The growth and prosperity of the metaverse depend critically on information security. In this section we consider metaverse data security with respect to: data used to create metaverse, Data transmission and sharing during metaverse implementation, data storage on network, and finally data provenance for metaverse. Figure 3 provides all these types of data.

4.1 Security for Metaverse Creation

Virtual items called "digital twin" are made to replicate real-world objects. These digital things can mimic the physical performance or behaviour of real-world assets, in addition to their physical look. Clones of real-world systems and items will be possible thanks to digital twins. The metaverse, where digital items would function similarly to physical ones, may be built on the foundation of digital twins. The metaverse's interactions can be leveraged to augment physical systems, leading to a large innovation path and better user experience. The metaverse must make sure that the digital twins built and used are original in order to safeguard them [31]. In order to safeguard the digital twins, the metaverse needs a trust-based information system. As users continue to participate, a significant quantity of transactional information will be finished on the blockchain because the metaverse will be constructed on a transactional infrastructure based on blockchain technology. Blockchain is a single, distributed chain in which cryptographic blocks are used to store the data [32]. Before adding the new record to the chain, a peer-to-peer network verifies the legitimacy of each new block (such as the creation of a new digital twin). A number of papers [32–34] suggest using blockchain-based solutions to safeguard digital twins in the metaverse. In [34], the authors put forth a blockchain-based system for digital twins to utilise to store health data electronically (such as biometric data). They can enable new types of markets in the digital ecosystems, such as non-fungible token (NFT) marketplaces, as we have seen with recent implementations [33]. The latter enables digital twin makers to use the blockchain to sell their digital twins as exclusive assets. Blockchain technologies, however, are evolving quickly. To ensure decentralisation and security in the metaverse, blockchain technology can be applied to the data storage system [32, 35].

AI technologies like the generative adversarial network (GAN) may provide dynamic game situations and context images of the highest quality, but they also present security risks like hostile and poisoned samples that are difficult for humans to identify. Through the use of virtual adversarial learning [36], adversarial representation learning [37], adversarial reinforcement learning [38], adversarial transfer learning [39], and other techniques, various attempts have been made to resist adversarial samples in the literature by using them as training data. These techniques can be useful for fending off threats to the metaverse's construction.

4.2 *Security for Storage Data on Metaverse*

Edge computing, which enables data processing and storage at the edge, would be a superior solution for security and privacy [40]. Instead of processing data in data centres or the cloud, edge computing uses hardware and software to process data locally, where it is received. As a result, networks may process data more quickly and effectively while using less bandwidth because there is no need to transfer data between servers and clients. Without requiring user private data uploads beyond local gradient adjustments, it can function on edge servers owned by end users and undertake extensive data mining over distributed clients. The metaverse's security and privacy can be improved using this technique (train at the edge and aggregate at the cloud).

4.3 *Security for Metaverse Data Sharing*

Applications for the metaverse are frequently multi-user, like online multiplayer games and remote teamwork. Ruth et al. [41].'s study of an AR content sharing control mechanism and implementation of a prototype on HoloLens are aimed at secure content sharing under multi-user AR applications. This enables AR content sharing among remote or collocated users with inbound and outbound control. The authors identify numerous ways of mapping AR contents into the actual world by carefully examining user design spaces on various AR apps. Lee et al. [42] describe three novel ad fraud concerns (i.e., blind spot monitoring, gaze and controller cursor-jacking, and manipulation of an auxiliary display) in content sharing for WebVR (a VR-based 3D virtual world using HTML canvases). According to user studies conducted on 82 people, success percentages can range from 88.23% to 100%. Additionally, [42] presents the AdCube defensive method, which uses sandboxing and visual confinement of 3D ad entities. Results from experiments demonstrate AdCube's defensive effectiveness for nine WebVR demo sites at a low system cost.

The aggregation and processing of massive data collected from human bodies and their surrounding environments are essential for the creation and rendering of avatars and virtual environments, in which users' sensitive information may be leaked [43]. In the metaverse, a lot of personally identifiable information collected from wearables (like HMDs) is transferred via wired and wireless communications. In order to train individualised avatar appearance models, for instance, private data from various users may be collected and stored in a single location in violation of the General Data Protection Regulation (GDPR) and other real-world laws [44]. Unauthorized parties or services should not have access to the confidentiality of these data. Even when information is transferred in confidence and communications are encrypted, adversaries may still get the raw data by listening in on a particular channel and even locate users using differential attacks [45] and advanced inference attacks [46].

4.4 Security for Data Provenance

Data provenance may make historical archives of a piece of UGC traceable, which is necessary to assess data quality, track down the source of the data, recreate the data generation process, and undertake audit trails to rapidly identify the individuals who are in charge of the data. UGC provenance data, including as its source, circulation, and intermediary processing details, are frequently kept in separate data silos (such as different blockchains) in the metaverse, making it challenging to monitor and follow in real time. The design of UGC provenance in the metaverse can learn from existing research on IoT data provenance. Satchidanandan and Kumar [47] create a dynamic watermarking technology that takes advantage of permanent patterns imprinted in the medium to identify malevolent sensors' or actuators' dishonest behaviour (such as signal tampering). In the metaverse, advanced watermarking techniques can also be used to safeguard intellectual property and prove ownership. Liang et al. [28] introduces ProvChain, a blockchain-based cloud file provenance architecture with three steps for gathering, storing, and verifying provenance data. Cloud storage reliability, user privacy, and source tamper resistance are all guaranteed by ProvChain. Kamal and Tariq [48] lightweight protocol for multi-hop IoT uses the RSS indicator of the transmitting IoT node to generate the distinctive connection fingerprint and enables data provenance in wireless communications.

5 Conclusions

Metaverse develops native technology that enables us to communicate with and share realities with the world around us. In the coming years, our virtual worlds will appear very different because of the application of developing technology as well as the steady growth and improvement of the ecosystem. We introduced the idea, traits, and enabling methods of the metaverse in this chapter. User data privacy and information security will face new issues as metaverse systems perceive and transact a growing amount of user data. We outlined the major issues with privacy and security in the metaverse in this chapter. We explore different threats and Security Countermeasures these threats. For the purpose of creating specialized security and privacy countermeasures for the metaverse, we have studied the current and possible solutions. We hope that this chapter will clarify how security and privacy are provided in metaverse applications and encourage more ground-breaking study in this rapidly developing field.

References

1. Njoku, J., Amaizu, G., Lee, J.M., Kim, D.-S.: Real-time deep learning-based scene recognition model for metaverse applications (2022)
2. Yassin, A.K.: The metaverse revolution and its impact on the future of advertising industry. J. Des. Sci. Appl. Arts **3**(2) (2022)
3. Wang, Y., Su, Z., Zhang, N., Xing, R., Liu, D., Luan, T.H., Shen, X.: A survey on metaverse: fundamentals, security, and privacy, communications surveys. Inst. Electr. Electron. Eng. (2022). https://doi.org/10.1109/comst.2022.3202047
4. Falchuk, B., Loeb, S., Neff, R.: The social metaverse: battle for privacy. IEEE Technol. Soc. Mag. **37**(2), 52–61 (2018)
5. Shang, J., Chen, S., Wu, J., Yin, S.: ARSpy: breaking location-based multi-player augmented reality application for user location tracking. IEEE Trans. Mob. Comput. **21**(2), 433–447 (2022)
6. Hu, P., Li, H., Fu, H., Cansever, D., Mohapatra, P.: Dynamic defense strategy against advanced persistent threat with insiders. In: IEEE Conference on Computer Communications (INFOCOM), pp. 747–755 (2015)
7. Huynh-The, T., Pham, Q.-V., Pham, X.-Q., Nguyen, T.T., Han, Z., Kim, D.-S.: Artificial intelligence for the metaverse: a survey. Eng. Appl. Artif. Intell. **117**, Part A, , 105581 (2023). ISSN 0952-1976. https://doi.org/10.1016/j.engappai.2022.105581
8. Zawish, M., Dharejo, F., Khowaja, S., Dev, K., Davy, S., Qureshi, N.M.F., Bellavista, P.: AI and 6G into the metaverse: fundamentals, challenges and future research trends (2022). https://doi.org/10.48550/arXiv.2208.10921.
9. Huang, Y., Qiao, X., Wang, H., Su, X., Dustdar, S., Zhang, P.: Multi-Player immersive communications and interactions in metaverse: challenges, architecture, and future direction, publisher (2022)

10. Dionisio, J.D.N., Burns, W.G., III, Gilbert, R.: 3D virtual worlds and the metaverse: current status and future possibilities. ACM Comput. Surv. (CSUR) **45**(3), 1–38 (2013)
11. ISO/IEC 23005 (MPEG-V) standards. https://mpeg.chiariglione.org/standards/mpeg-v. Accessed 20 Sep. 2021
12. IEEE 2888 standards. https://sagroups.ieee.org/2888/. Accessed 20 Dec. 2021
13. Shen, X., Miao, C.: A full dive into realizing the edge-enabled metaverse: visions, enabling technologies, and challenges. arXiv:2203.05471 (2022)
14. Jayasinghe, U., Lee, G.M., Um, T.-W., Shi, Q.: Machine learning based trust computational model for IoT services. IEEE Trans. Sustain. Comput. **4**(1), 39–52 (2019)
15. Kosinski, M., Stillwell, D., Graepel, T.: Private traits and attributes are predictable from digital records of human behavior. Proc. Natl. Acad. Sci. **110**(15), 5802–5805 (2013)
16. Laperdrix, P., Bielova, N., Baudry, B., Avoine, G.: Browser fingerprinting: a survey. ACM Trans. Web **14**(2) (2020)
17. Leenes, R.: Privacy in the metaverse. In: IFIP International Summer School on the Future of Identity in the Information Society, pp. 95–112. Springer (2007)
18. Boddington, G.: The internet of bodies—alive, connected and collective: the virtual physical future of our bodies and our senses. Ai & Soc. 1–17 (2021)
19. Ouda, O., Tsumura, N., Nakaguchi, T.: Bioencoding: a reliable tokenless cancelable biometrics scheme for protecting iriscodes. IEICE Trans. Inf. Syst. **93**(7), 1878–1888 (2010)
20. Kürtünlüoğlu, P., Akdik, B., Karaarslan, E.: Security of virtual reality authentication methods in metaverse: an overview (2022)
21. Grover, V.: Technology: a tangible threat to our privacy. Res. Journali's J. Sociol. **3**, 1–9 (2015)
22. Mathis, F., Fawaz, HI, Khamis, M.: Knowledge-driven biometric authentication in virtual reality. In: Extended Abstracts of the 2020 CHI Conference on Human Factors in Computing Systems, pp. 1–10 (2020)
23. Grubert, J., Langlotz, T., Zollmann, S., Regenbrecht, H.: Towards pervasive augmented reality: context-awareness in augmented reality. IEEE Trans. Visual Comput. Graph. **23**(6), 1706–1724 (2017)
24. Su, Z., Wang, Y., Xu, Q., Zhang, N.: LVBS: Lightweight vehicular blockchain for secure data sharing in disaster rescue. IEEE Trans. Dependable Secur. Comput. **19**(1), 19–32 (2022)
25. https://krontech.com/cybersecurity-risks-that-use-metaverse-as-a-tool
26. Liang, G., Weller, S.R., Zhao, J., Luo, F., Dong, Z.Y.: The 2015 Ukraine blackout: Implications for false data injection attacks. IEEE Trans. Power Syst. **32**(4), 3317–3318 (2017)
27. Lee, L.-H., Braud, T., Zhou, P., Wang, L., Xu, D., Lin, Z., Kumar, A., Bermejo, C., Hui, P.: All one needs to know about metaverse: a complete survey on technological singularity, virtual ecosystem, and research agenda (2021). https://doi.org/10.13140/RG.2.2.11200.05124/8
28. Liang, X., Shetty, S., Tosh, D., Kamhoua, C., Kwiat, K., Njilla, L.: Provchain: a blockchain-based data provenance architecture in cloud environment with enhanced privacy and availability. In: IEEE/ACM International Symposium on Cluster, Cloud and Grid Computing (CCGRID), pp. 468–477 (2017)
29. De Ree, M., Mantas, G., Radwan, A., Mumtaz, S., Rodriguez, J., Otung, I.E.: Key management for beyond 5G mobile small cells: a survey. IEEE Access **7**, 59 200–59 236 (2019)
30. Duan, H., Li, J., Fan, S., Lin, Z., Wu, X., Cai, W.: Metaverse for social good: a university campus prototype. In: ACM International Conference on Multimedia (MM), pp. 153–161 (2021)
31. Rasheed, A., San, O., Kvamsdal, T.: Digital twin: values, challenges and enablers from a modeling perspective. IEEE Access **8**, 21980–22012 (2020)
32. Reyna, A., Mart´ın, C., Chen, J., Soler, E., D´ıaz, M.: On blockchain and its integration with iot. Challenges and opportunities. Futur. Gener. Comput. Syst. **88**, 173–190 (2018)
33. Omar, A.S., Basir, O.: Capability-based non-fungible tokens approach for a decentralized aaa framework in iot. In: Blockchain Cybersecurity, Trust and Privacy, pp. 7–31. Springer (2020)
34. Chen, L., Lee, W.-K., Chang, C.-C., Choo, K.-K.R., Zhang, N.: Blockchain based searchable encryption for electronic health record sharing. Futur. Gener. Comput. Syst. 95:420–429 (2019)
35. Acquisti, A., Brandimarte, L., Loewenstein, G.: Privacy and human behavior in the age of information. Science **347**(6221), 509–514 (2015)

36. Miyato, T., Maeda, S.-I., Koyama, M., Ishii, S.: Virtual adversarial training: A regularization method for supervised and semi-supervised learning. IEEE Trans. Pattern Anal. Mach. Intell. **41**(8), 1979–1993 (2019)
37. Mai, S., Hu, H., Xing, S.: Modality to modality translation: an adversarial representation learning and graph fusion network for multimodal fusion. In: AAAI, pp. 1–9 (2019)
38. Sun, J., Zhang, T., Xie, X., Ma, L., Liu, Y.: Stealthy and efficient adversarial attacks against deep reinforcement learning. In: AAAI, pp. 1–9 (2020)
39. Zheng, H., Zhang, Z., Gu, J., Lee, H., Prakash, A.: Efficient adversarial training with transferable adversarial examples. In: IEEE/CVF Conference on Computer Vision and Pattern Recognition (CVPR), pp. 1–10 (2020)
40. Mollah, M.B., Azad, MdA.K., Vasilakos, A.: Security and privacy challenges in mobile cloud computing: Survey and way ahead. J. Netw. Comput. Appl. **84**, 38–54 (2017)
41. Ruth, K., Kohno, T., Roesner, F.: Secure multi-user content sharing for augmented reality applications. In: 28th USENIX Security Symposium (USENIX Security 19), pp. 141–158 (2019)
42. Lee, H., Lee, J., Kim, D., Jana, S., Shin, I., Son, S.: AdCube: WebVR ad fraud and practical confinement of Third-Party ads. In: 30th USENIX Security Symposium (USENIX Security 21), pp. 2543–2560 (2021)
43. Li, X., He, J., Vijayakumar, P., Zhang, X., Chang, V.: A verifiable privacy-preserving machine learning prediction scheme for edge-enhanced HCPSs. IEEE Trans. Industr. Inf. (2021). https://doi.org/10.1109/TII.2021.3110808
44. General Data Protection Regulation (GDPR). https://gdpr-info.eu/. Accessed 2 March 2022.
45. Wei, J., Li, J., Lin, Y., Zhang, J.: LDP-based social content protection for trending topic recommendation. IEEE Internet Things J. **8**(6), 4353–4372 (2021)
46. Wasserkrug, S., Gal, A., Etzion, O.: Inference of security hazards from event composition based on incomplete or uncertain information. IEEE Trans. Knowl. Data Eng. **20**(8), 1111–1114 (2008)
47. Satchidanandan, B., Kumar, P.R.: Dynamic watermarking: active defense of networked cyber–physical systems. Proc. IEEE **105**(2), 219–240 (2017)
48. Kamal, M., Tariq, S.: Light-weight security and data provenance for multi-hop internet of things. IEEE Access **6**, 34 439–34 448 (2018)

Printed in the United States
by Baker & Taylor Publisher Services